CHEMICAL AND FUNCTIONAL PROPERTIES OF FOOD COMPONENTS

HOW TO ORDER THIS BOOK

BY PHONE: 800-233-9936 or 717-291-5609, 8AM–5PM Eastern Time

BY FAX: 717-295-4538

BY MAIL: Order Department
Technomic Publishing Company, Inc.
851 New Holland Avenue, Box 3535
Lancaster, PA 17604, U.S.A.

BY CREDIT CARD: American Express, VISA, MasterCard

BY WWW SITE: http://www.techpub.com

PERMISSION TO PHOTOCOPY–POLICY STATEMENT

Authorization to photocopy items for internal or personal use, or the internal or personal use of spe-
cific clients, is granted by Technomic Publishing Co., Inc. provided that the base fee of US $3.00 per
copy, plus US $ .25 per page is paid directly to Copyright Clearance Center, 222 Rosewood Drive,
Danvers, MA 01923, USA. For those organizations that have been granted a photocopy license by
CCC, a separate system of payment has been arranged. The fee code for users of the Transactional
Reporting Service is 1-56676/97 $5.00 + $ .25.

CHEMICAL AND FUNCTIONAL PROPERTIES OF FOOD COMPONENTS

Edited by

ZDZISŁAW E. SIKORSKI, Ph.D., D.Sc.

Professor of Food Science
Department of Food Preservation
Technical University of Gdansk, Poland

TECHNOMIC
PUBLISHING CO., INC.
LANCASTER · BASEL

Chemical and Functional Properties of Food Components

a **TECHNOMIC**® publication

Published in the Western Hemisphere by
Technomic Publishing Company, Inc.
851 New Holland Avenue, Box 3535
Lancaster, Pennsylvania 17604 U.S.A.

Distributed in the Rest of the World by
Technomic Publishing AG
Missionsstrasse 44
CH-4055 Basel, Switzerland

Printed in the United States of America
10 9 8 7 6 5 4 3 2 1

Main entry under title:
    Chemical and Functional Properties of Food Components

A Technomic Publishing Company book
Bibliography: p.
Includes index p. 285

Library of Congress Catalog Card No. 96-61440
ISBN No. 1-56676-464-5

Table of Contents

v

Preface

EVERYBODY expects food to be palatable, nutritious, and wholesome. These attributes are affected by the composition of the food, as well as by the properties and interactions of the constituents. They can be controlled by the recipe of the formulation, the conditions of storage and processing, and the enzymatic or chemical modifications of the components, mainly lipids, proteins, saccharides, and vitamins. The effectiveness of control depends on the understanding of the chemical and functional properties of the food components. Thus, food chemistry is a must for food technologists and for other technical personnel involved in food handling.

Food chemistry is a broad and rapidly developing science. The easy, effecient way to prepare an up-to-date and concise presentation of the main topics in this field is to do it in cooperation with several specialists. I have been lucky to find the understanding of colleagues from Australia, the Netherlands, Poland, Taiwan, and the United States who have willingly contributed to this volume, sharing their knowledge and experience. I am thankful for their expert input. My thanks are also due to all persons who have granted permission for using their previously published material.

I am glad to be able to dedicate this volume to the memory of Dr. Fred E. Deatherage, former Professor and Head, Department of Agricultural Biochemistry, Ohio State University. My interest in food chemistry and in the functional properties of food proteins began during my postdoctoral training in his department, at that time a recognized center of meat biochemistry, in 1964–1965. Professor Deatherage was renown for his worldwide cooperation in meat research. Over the past 30 years, I have studied food chemistry problems in different countries. My international connections have made it possible to assemble the team for preparing this book.

My absence from home during periods of research abroad and the time-consuming hobby of writing food science books has been possible only because I have the understanding of my wife Krystyna.

ZDZISŁAW E. SIKORSKI

Contributors

AGNIESZKA BARTOSZEK, Ph.D., Department of Biochemistry, Technical University of Gdańsk, Poland

BARBARA CYBULSKA, Ph.D., Department of Biochemistry, Technical University of Gdańsk, Poland

PETER DOE, Ph.D., Professor, Department of Engineering, University of Tasmania, Launceston, Australia

TADEUSZ S. MATUSZEK, Ph.D. Department of Mechanical Engineering, Technical University of Gdańsk, Poland

JULIE MILLER JONES, Ph.D., Professor, Department of Home Economics, College of St. Catherine, St. Paul, Minnesota, USA

MICHAŁ NABRZYSKI, Ph.D., D.Sc., Professor, Department of Bromatology, Medical Academy of Gdańsk, Poland

BONNIE SUN PAN, Ph.D., Professor, Department of Marine Food Science, National Taiwan Ocean University, Keelung, Taiwan, Republic of China

ADRIAAN RUITER, Ph.D., Professor, Department of the Science of Food of Animal Origin, Utrecht University, the Netherlands

JAN SAWICKI, Ph.D., Department of Chemistry and Technology of Fats, Technical University of Gdańsk, Poland

ZDZISŁAW E. SIKORSKI, Ph.D., D.Sc., Professor, Department of Food Preservation, Technical University of Gdańsk, Poland

BARTŁOMIEJ ŚLUSARSKI, M.Sc., Department of Technology of Lipids and Detergents, Technical University of Gdańsk, Poland

PIOTR TOMASIK, Ph.D., D.Sc., Professor, Department of Chemistry, Academy of Agriculture, Kraków, Poland

ALPHONS G. J. VORAGEN, Ph.D., Professor, Department of Food Science, Utrecht University, the Netherlands

JADWIGA WILSKA-JESZKA, Ph.D., D.Sc., Professor emeritus, Institute of Technical Biochemistry, Technical University of Łódź, Poland

CHUNG-MAY WU, Ph.D., Senior Scientist, Food Research and Development Institute, Hsinchu, Taiwan, Republic of China

CHAPTER 1

Food Components and Their Role in Food Quality

ZDZISŁAW E. SIKORSKI

THE MAIN FOOD COMPONENTS

INTRODUCTION

OUR foods are derived from plant material, carcasses of animals, and single-cell organisms. They are composed mainly of water, saccharides, proteins, lipids, and minerals (Table 1.1). These main components serve as nutrients by supplying the human body with the necessary building materials and source of energy, as well as elements and compounds indispensable for metabolism. Foods also contain a host of other constituents present in smaller quantities, especially various nonprotein nitrogenous compounds, vitamins, colorants, flavor compounds, and functional additives. Many of the minor components present originally in the food are essential, e.g., the vitamins; some can be utilized by our body, e.g., amino acids, while others are useless or even harmful if they occur in excessive amounts. A variety of compounds is added intentionally during processing.

THE ROLE IN FOOD RAW MATERIALS

Polysaccharides, proteins, and lipids are components of different structures of the plant and animal tissues used for food. The structures made of these materials are responsible for the form and tensile strength of the tissues and create the necessary conditions for the metabolic processes. Compartmentation resulting from these structures plays a crucial biological role in the food organisms. Some saccharides, proteins, and lipids are stored for reserve purposes. Other constituents are either bound to the

1

TABLE 1.1. **The Main Components in Typical Foods.**

Water	Saccharides	Proteins	Lipids	Minerals	Vitamins
Juices	Saccharose	Soybean	Vegetable	Vegetables	Vegetables
Fruits	Honey	Beans	Oil	Fruits	Fruits
Milk	Cereals	Meat	Lard	Meat	Fish liver
Vegetables	Chocolate	Fish	Butter	Fish, including	Meat
	Potato	Wheat	Chocolate	bones	Cereals
Jellies	Cassava	Cheese	Nuts	Dairy products	Milk
Lean fish	Fruits	Egg	Egg yolk	Nuts	
			Pork		

structural or metabolically active compounds or are distributed in soluble form in the cells and tissue fluids.

The contents of water in different foods range from a few percent in dried commodities to about 15% in grains, 75% in meat, and about 90% in fruits and vegetables. Most of the water is immobilized in the plant and animal tissues by the structural elements and various solutes, contributes to buttressing the conformation of the polymers, and interacts in metabolic processes.

Saccharides are present in food raw materials in quantities ranging from about 1% in meats and fish, about 4.5% in milk and 18% in potatoes to about 70% in cereal grains. Polysaccharides participate in the formation of structures in plants. They are also stored in plants as starch and in muscles as glycogen. Other saccharides are dissolved in tissue fluids or perform different biological functions, e.g., in free nucleotides or as components of nucleic acids or being bound to proteins and lipids.

The protein contents in foods are given mainly as crude protein, i.e., generally as $N \times 6.25$. The true nitrogen-to-protein conversion factors are, however, different for the proteins in various foods because they depend on the amino acid composition. Furthermore, the total N consists of protein N and of N contained in different nonprotein compounds, e.g., free peptides and amino acids, nucleic acids and their degradation products, amines, betains, urea, vitamins, and alcaloids. In some foods, the nonprotein N may constitute up to 30% of total N. In many of these compounds, the C:N ratio is similar to the average in amino acids; however, the N content in urea, being 47%, is exceptionally high. Most of the nonprotein nitrogen compounds can be utilized by the organism as a source of nitrogen.

The average conversion factor estimated by Sosulski and Imafidon (1990) for twenty-three various food products is 5.68 and, for different classes of foods, is in the range of 5.14–6.61 (Table 1.2). A common N:P conversion factor of 5.70 for use in respect to blended foods or diets has been recommended in Sosulski and Imafidon (1990).

Proteins make up about 1% of the weight of fruits and 2% of potatoes, 12% of eggs, 12–22% of wheat, about 20% of meat, and up to 25–40% of different beans. They serve as the building material of muscles and other animal tissues and play crucial metabolic roles, not only as enzymes, in plants and animals. During their development, cereal grain and legume seeds deposit large quantities of storage proteins in granules known also as protein bodies. In soybeans, these proteins constitute 60–70% of the total protein content, and the granules are 80% proteins.

Lipids constitute from below 1% of the weight of fruits, vegetables, and lean fish; 3.5% of milk; 6% of beef; 32% of egg yolk, and 85% of butter. The lipids contained in the food raw materials in low quantities are mainly involved in the structure of protein-phospholipid membranes and perform metabolic functions. In fatty commodities, the majority of lipids is stored as depot fat in the form of triacylglycerols. The lipid fraction also harbors different vitamins and pigments that are crucial for the metabolism.

FACTORS AFFECTING FOOD COMPOSITION

The contents of different components in food raw materials depend on the species and variety of the animal and plant crop, on the conditions of cultivation and harvesting of the plants, and on the feeding and age of the farm animals or the season and fishing method of seafoods. The food industry, by establishing quality requirements for raw materials, can induce the producers to control within limits the contents of the main components in their crops, e.g., saccharose in sugar beets, starch in potatoes, fat in various meat cuts, certain desirable or unwanted fatty acids in oil crops, pigments in fruits and vegetables, or protein in wheat and barley. Contamination of the raw material with organic and inorganic pollutants can be controlled, e.g., by observing recommended agricultural procedures in using fertilizers, herbicides, and insecticides, by seasonally restricting

TABLE 1.2. The Nitrogen to Protein Conversion Factors in Foods.

Product	Factor	Product	Factor
Dairy products	6.02–6.15	Potato	5.18
Egg	5.73	Leafy vegetables	5.14–5.30
Meat and fish	5.72–5.82	Fruits	5.18
Cereals and legumes	5.40–5.93	Microbial biomass	5.78–6.61

Source: Data from Sosulski and Imafidan (1990).

certain fishing areas to avoid marine toxins, or by limiting the size of predatory fish like swordfish, tuna, or sharks to reduce the risk of too high a content of mercury in the flesh.

The composition of processed foods depends on the applied recipe and on changes taking place due to processing and storage. Such changes are mainly brought about by enzymes, active forms of oxygen, heating, chemical treatment, and processing at low or high pH. Examples of such changes are

- leaching of soluble components during washing or blanching
- drip formation after thawing or due to cooking
- loss of moisture and volatiles due to evaporation
- absorption of desirable or harmful compounds, e.g., during salting, pickling, or smoking
- generation of new nutritionally objectionable or required compounds due to interactions of reactive groups induced by heating
- autooxidation

THE QUALITY OF FOODS

The quality of a food product, i.e., the characteristic properties that determine the degree of excellence, is a sum of the attributes contributing to the satisfaction of the consumer with the product. The composition and the chemical nature of the food components affect all aspects of food quality. The total quality reflects at least the following attributes:

- compatibility with the food law regulations and standards, local or international, that regard mainly hygienic requirements, contents of contaminants and additives, proportions of main components, and packaging
- nutritional quality, i.e., the contents of nutritionally desirable components, mainly proteins, essential amino acids, essential fatty acids, vitamins, fiber, and microelements
- safety aspects affected by the contents of compounds that may constitute health hazards for the consumers and affect the digestibility and nutritional use of the food, e.g., heavy metals, toxins of different origin, pathogenic microorganisms, and inhibitors of digestive enzymes
- sensory quality, i.e., the color, size, form, flavor, taste, and rheological properties obviously affected by the chemical composition of the product
- convenience aspects that are reflected by the size and ease of opening/reclosing of the container, suitability of the product for

immediate use or for different types of thermal treatment, ease of portioning or spreading, storage requirements, etc.
* ecological aspects regarding suitability for recycling of the packaging material and pollution hazards

THE FUNCTIONAL PROPERTIES OF FOOD COMPONENTS

The term *functional properties* has recently evolved to have a broad range of meanings. That corresponding to the term *technological properties* implies that the given component present in optimum concentration, subjected to processing at optimum parameters, contributes to the desirable sensory properties of the product, usually by interacting with other food constituents. The effects result due to hydrogen bonds, hydrophobic interactions, ionic forces, and covalent bonding. Thus, the functional properties of food components are affected by the number of accessible reactive groups and on the exposure of hydrophobic areas in the given medium. Therefore, the functional properties displayed in a medium of given water activity and pH and in the given range of temperature can be, to a large extent, predicted from the structure of the respective saccharides, proteins, and lipids. They can also be improved by appropriate, intentional enzymatic or chemical modifications of the molecules, mainly those that affect the size, charge density, or hydrophilic/hydrophobic character of the compounds or by changes in the environment, regarding both the solvent and other solutes.

The functional properties of food components make it possible to manufacture products of desirable quality. Thus, pectins contribute to the characteristic texture of ripe apples and make perfect jellies. Various other polysaccharides are good thickening and gelling agents at different ranges of acidity. Some starches are resistant to retrogradation, thereby retarding staling of bread. Alginates form protective, unfrozen gels on the surface of frozen products. Fructose retards moisture loss from biscuits. Mono- and diacylglycerols, phospholipids, and proteins are used for emulsifying lipids and stabilizing food emulsions; antifreeze proteins decrease ice formation in different products, and gluten plays a major role in producing the characteristic texture of wheat bread.

Technologically required functional effects can also be achieved by intentionally employing various food additives, i.e., food colors, sweeteners, and a host of other compounds, which are not regarded as foodstuffs per se but are used to modify the rheological properties or acidity, increase the color stability or shelf life, act as humectants or flavor enhancers, etc. (Rutkowski et al., 1993). These compounds are exhaustively treated in Chapter 10.

During the recent decade, the term *functional* has also been given to a large group of products, also termed designer foods, pharmafoods, nutraceuticals, or foods of specific health use, which are regarded as health-enhancing. These foods, mainly drinks, meals, confectionery, ice cream, and salad dressings, contain various ingredients, e.g., oligosaccharides, sugar alcohols, or choline, which are claimed to have special physiological functions like neutralizing harmful compounds in the body and promoting recovery and general good health (Goldberg, 1994).

THE ROLE OF CHEMISTRY AND PROCESSING FACTORS

INTRODUCTION

The chemical nature of food components is of crucial importance for all aspects of food quality. It decides upon the nutritional value of the product, its sensory attractiveness, development of desirable or deteriorative changes due to interactions with other food constituents and to processing, and susceptibility/resistance to spoilage during storage. Food components that contain reactive groups, many of them essential for the quality of the products, are generally labile and easily undergo different changes, especially when treated at elevated temperature or in conditions promoting the generation of active species of oxygen.

EFFECT ON NUTRITIONAL VALUE

The nutritional value of foods depends primarily on the contents of nutrients and nutritionally objectionable components in the products.

Processing may increase the biological value of the food by inducing chemical changes enhancing the digestibility of the components or by inactivating undesirable compounds, e.g., toxins or enzymes catalyzing the generation of toxic agents from harmless precursors. A typical case of the latter is intentional thermal inactivation of myrosinase, the enzyme involved in hydrolysis of glucosinolanes. Inactivation of the enzyme arrests the reactions that lead to the formation of goitrogenic products in oilseeds of *Cruciferae*. Several other examples of such improvements of the biological quality of foods are given in the following chapters of this book.

There are, however, also nutritionally undesirable side effects of processing, i.e., destruction of essential food components as a result of heating, chemical treatment, and autoxidation. Best known is the partial thermal decomposition of vitamins, especially thiamine, loss of available lysine and sulphur-containing amino acids, or generation of unnatural compounds, e.g., lanthionine and other amino acids or position isomers of fatty acids not present originally in foods. Thanks to the unprecedented development of analytical chemistry, it became possible, by applying efficient procedures of enrichment and separation combined with the use of highly selective and sensitive detectors, to determine various products of chemical reactions in foods even in very low concentrations. In recent years, new evidence of side effects has been accumulated in respect to the conventional chemical processing of oils and fats. Commercial hydrogenation of oils brings about not only the intended saturation of selected double bonds in the fatty acids and thereby the required change in the rheological properties of the oil, but also results in generation of a large number of *trans-trans* and *cis-trans* isomers, which are absent in the virgin oils (Sto-Żyhwo, 1995).

EFFECT ON SENSORY QUALITY

Much of the desirable sensory quality of foods stems from the properties of the raw material, e.g., the color, flavor, taste, and texture of fruits, vegetables, and juices or the taste of nuts and milk. These properties are, in many cases, carried through to the final products.

In other commodities, the characteristic quality attributes are generated in processing, e.g., the texture of bread due to interactions of proteins, lipids, and saccharides with each other and with various gases; the bouquet of wine due to fermentation of saccharides and a number of other biochemical and chemical reactions; the delicious color, flavor, texture, and taste of smoked salmon due to enzymatic changes in the tissues and the effect of salt and smoke; the flavor, texture, and taste of cheese generated due to fermentation and ripening; and the appealing color and flavor of different fried products due to reactions of saccharides and amino acids.

The sensory quality of foods is related to the contents of many chemically labile components. These components, however, like most nutritionally essential compounds, are also prone to deteriorative changes in conditions of severe heat treatment or application of considerably high doses of chemical agents, e.g., acetic acid or salt, which are often required to ensure safety of the products. Thus, loss of sensory quality takes place, e.g., in oversterilized meat products due to degradation of sulphur-containing amino acids and development of off-flavor; toughening of texture of overpasteurized ham or shellfish due to excessive shrinkage of the tissues and drip; deterioration of the texture and arresting of ripening in herring preserved at too high a concentration of salt.

To apply optimum parameters of processing that ensure the retention of the desirable sensory properties of the raw material, lead to the development of the intended attributes of the product, and avoid the deterioration of sensory quality, the chemical nature of the effect of processing on the components of foods must be studied. The eager food technology student can find all the necessary information in the two excellent textbooks on food chemistry published by Belitz and Grosch (1992) and by Fennema (1996) and in numerous books on food lipids, proteins, and saccharides, as well as in the current international journals.

REFERENCES

Belitz, H. D. and Grosch, W. 1992. *Lehrbuch der Lebensmittelchemie.* Fourth edition. Berlin: Springer Verlag.

Fennema, O. R., ed. 1996. *Food Chemistry.* Third edition. New York and Basel: Marcel Dekker.

Goldberg, I. 1994. *Functional Foods: Designer Foods, Pharmafoods, Nutraceuticals.* New York: Chapman and Hall.

Rutkowski, A., Gwiazda, S. and Dabrowski, K. 1993. *Functional Food Additives.* Katowice: Agro & Food Technology (in Polish).

Sosulski, F. W. and Imafidon, G. I. 1990. "Amino acid composition and nitrogen-to-protein conversion factors for animal and plant foods," *J. Agric. Food Chem.* 38:1351–1356.

Stołyhwo, A. 1995. "HPLC and HR-GC in analysing the composition of natural and modified fats, margarines, and mothers' milk." *Proceedings of the Merck Conference on Methods of Food Analysis in Perspective of the Requirements of the European Union.* November 1995, Warsaw, pp. 23–39 (in Polish).

CHAPTER 2

Water and Food Quality

BARBARA E. CYBULSKA
PETER E. DOE

INTRODUCTION

WATER is the most popular and most important chemical compound on our planet. It is a major chemical constituent of the earth's surface and is the only substance that is abundant in solid, liquid, and gaseous form. Because it is ubiquitous, it seems to be a mild and inert substance. In fact, it is a very reactive compound characterized by unique physical and chemical properties, which make it very different from other popular liquids. The peculiar water properties determine the nature of the physical and biological world.

Water is the major component of all living organisms. It constitutes 60% or more of the weight of most living things, and it pervades all portions of every cell. It existed on our planet long before appearance of any form of life. The evolution of life was doubtlessly shaped by physical and chemical properties of the aqueous environment. All aspects of living cells' structure and function seem to be adapted to water unique properties.

Water is the universal solvent and dispersing agent, as well as a very reactive chemical compound. Biologically active structures of biomacromolecules are spontaneously formed only in the aqueous media. Intercellular water is not only a medium in which structural arrangement and all metabolic processes occur, but it is an active partner of the molecular interactions and also participates directly in many biochemical reactions as a substrate or as a product. The high heat capacity allows water to act as a heat buffer in all organisms. Regulation of water contents is important in the maintenance of homeostasis in all living systems.

Only 0.003% of all reserves of sweet water participate in its continuous circulation between the atmosphere and the hydrosphere. The remaining part is confined in the Antarctic ice. The geography of water availability had determined, to a large degree, the vegetation, food supply, and habitation in the various areas of the world. Bangladesh has one of the world's highest population densities—made possible through the regular flooding of the Ganges River and the rich silts it deposits in its wake. In Bangladesh, the staple food—rice—grows abundantly and is readily distributed. In other societies, the food must be transported long distances or kept over winter. Stability, wholesomeness, and shelf life are significant features of such foods. These features are, to a large degree, influenced by the water content of the food. Dried foods were originally developed to overcome the constraints of time and distance before consumption. Canned and frozen foods developed next. The form, quantity, and quality of water within food plays a vital part in their effectiveness.

STRUCTURE AND PROPERTIES OF WATER

THE WATER MOLECULE

Water is a familiar material but has been described as the most anomalous of chemical compounds. While its chemical composition, HOH or H_2O, is universally known, the simplicity of its formula belies the complexity of its behavior. Its physical and chemical properties are very different from compounds of similar complexity, such as HF and H_2S. To understand the reasons for water's unusual properties, it is necessary to examine its molecular structure in some detail.

Although a water molecule is electrically neutral as a whole, it has a dipolar character. The high polarity of water is caused by the direction of the $H-O-H$ bond angle, which is $104.5°$, and by an asymmetrical distribution of electrons within the molecule. In a single water molecule, each hydrogen atom shares an electron pair with the oxygen atom in a stable covalent bond. However, the sharing of electrons between H and O is unequal because the more electronegative oxygen atom tends to draw electrons away from the hydrogen nuclei. The electrons are more often in the vicinity of the oxygen atom than of the hydrogen atom. The result of this unequal electron sharing is the existence of two electric dipoles in the molecule, one along each of the $H-O$ bonds. The oxygen atom bears a partial negative charge δ^- and each hydrogen a partial positive charge δ^+. Since the molecule is not linear, $H-O-H$ has a dipole moment (Figure 2.1). Due to that, water molecules can interact through electrostatic attraction between the oxygen atom of one water molecule and the hydrogen of another. The nearly tetrahedral arrangement of the orbital about the oxygen atom allows each water molecule to form hydrogen bonds with four

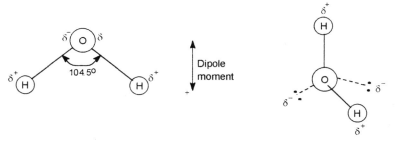

Figure 2.1 The water molecule as an electric dipole.

of its neighbors (Figure 2.2) (Lehninger et al., 1993; Nossal and Lecar, 1991).

HYDROGEN BONDS

Such interactions, which arise because the electrons on one molecule can be partially shared with the hydrogen on another, are known as hydrogen bonds. The H_2O molecule, which contains two hydrogen atoms and one oxygen atom in a nonlinear arrangement, is ideally suited to engage in hydrogen bonding. It can act both as a donor and as an acceptor of hydrogens. An individual, isolated hydrogen bond is very labile. It is longer and weaker than a covalent O−H bond (Figure 2.3). The hydrogen bond's energy, i.e., the energy required to break the bond, is about 20 kJ/mol. These bonds are intermediate between those of weak van der Waals interactions (about 1.2 kJ/mol) and those of covalent bonds (460 kJ/mol). Hydrogen bonds are highly directional; they are stronger when the hydrogen and the two atoms that share it are in a straight line (Figure 2.4).

Hydrogen bonds are not unique to water. They are formed between water and different chemical structures, as well as between other molecules or even within a molecule. They are formed wherever an elec-

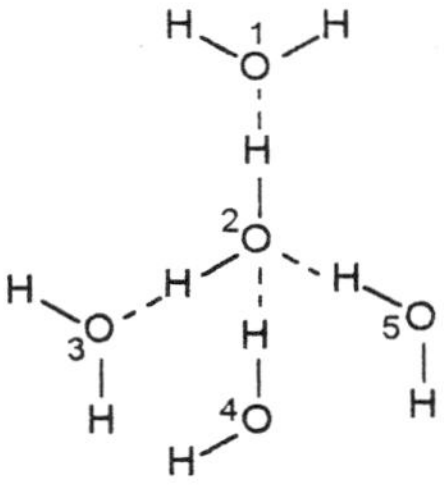

Figure 2.2 Tetrahedral hydrogen bonding of five water molecules.

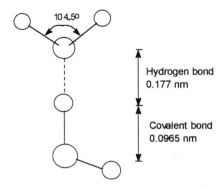

Figure 2.3 Two water molecules connected by hydrogen bonds.

tronegative atom (oxygen or nitrogen) comes in close proximity to a hydrogen covalently bonded to another electronegative atom. Some representative hydrogen bonds of biological importance are shown in Figure 2.5.

Intra- and intermolecular hydrogen bonding occurs extensively in biological macromolecules. A large number of the hydrogen bonds and its directionality confers very precise three-dimensional structures upon proteins and nucleic acids.

PROPERTIES OF BULK WATER

The key to understanding water structure in solid and liquid form lies in the concept and nature of the hydrogen bonds. In the crystal of ordinary hexagonal ice (Figure 2.6), each molecule forms four hydrogen bonds with its nearest neighbors. Each HOH acts as a hydrogen donor to two of the four water molecules and as a hydrogen acceptor from the remaining two. These four hydrogen bonds are spatially arranged according to the tetrahedral symmetry.

The crystal lattice of ice occupies more space than the same number of H_2O molecules in liquid water. The density of solid water is thus less than

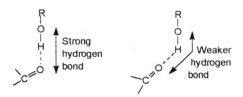

Figure 2.4 Directionality of the hydrogen bonds.

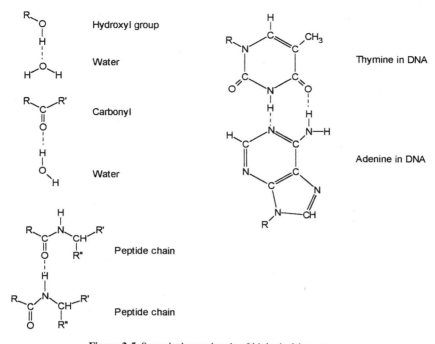

Figure 2.5 Some hydrogen bonds of biological importance.

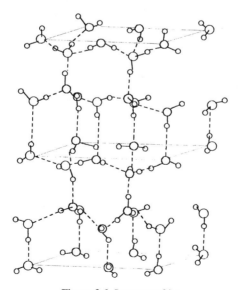

Figure 2.6 Structure of ice.

13

that of liquid water, whereas simple logic would have the more tightly bound solid structure more dense than its liquid. One explanation for ice being lighter than water at 0°C proposes a reforming of intermolecular bonds as ice melts so that, on average, a molecule is bound to more than four of its neighbors, thus increasing its density. But as the temperature of liquid water increases, the intermolecular distances also increase, giving a lower density. These two opposite effects explain the fact that liquid water has a maximum density at a temperature of 4°C. At any given instant in liquid water at room temperature, each water molecule forms hydrogen bonds with an average of 3.4 other water molecules (Lehninger et al., 1993). The average translational and rotational kinetic energies of a water molecule are approximately 7 kJ/mol, the same order as that required to break hydrogen bonds; therefore, hydrogen bonds are in a continuous state of flux, breaking and reforming with high frequency on a picosecond time scale. A similar dynamic process occurs in aqueous media with substances that are capable of forming hydrogen bonds.

At 100°C, liquid water still contains a significant number of hydrogen bonds, and even in water vapor, there is strong attraction between water molecules (Stillinger, 1980). The very large number of hydrogen bonds between molecules confers great internal cohesion on liquid water. This feature provides a logical explanation for many of its unusual properties. For example, its large values for heat capacity, melting point, boiling point, surface tension, and heat of various phase transitions are all related to the extra energy needed to break intermolecular hydrogen bonds.

That liquid water has structure is an old and well accepted idea; however, there is no consensus among physical chemists as to the molecular architecture of the hydrogen bond's network in the liquid state. The available measurements on liquid water do not lead to a clear picture of liquid water structure. It seems that the majority of hydrogen bonds survive the melting process, but obviously, rearrangements of molecules occur. The replacement of crystal rigidity by fluidity gives molecules more freedom to diffuse about and to change their orientation. Any molecular theory for liquid water must take into account changes in the topology and geometry of the hydrogen bond network induced by the melting process.

Many models have been proposed, but none adequately has explained all properties of the liquid water. "Iceberg" models postulated that liquid water contains disconnected fragments of ice suspended in a sea of unbounded water molecules.

The most popular, the so-called "flickering clusters" model, suggests that liquid water is highly organized on a local basis: the hydrogen bonds break and reform spontaneously, creating and destroying transient structural domains (Figure 2.7) (Lehninger et al., 1993). However, because the half-life of any hydrogen bond is less than a nanosecond, the existence of

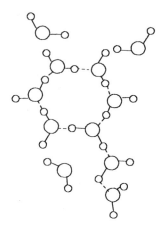

Figure 2.7 "Flickering clusters" of H_2O molecules in bulk water.

these clusters has statistical validity only; even this has been questioned by some authors who consider water to be a continuous polymer.

Experimental evidence obtained by X-rays and neutron diffraction strongly supports the persistence of tetrahedral hydrogen-bond order in the liquid water, but with substantial disorder present.

Stillinger (1980) created a qualitatively "water-like" structure by computer simulation. The view that does emerge from these results is the following: liquid water consists of a macroscopically connected, random network of hydrogen bonds. This network has a local preference for tetrahedral geometry, but it contains a large proportion of strained and broken bonds, which are continually undergoing topological reformation. The properties of water arise from the competition between relatively bulky ways of connecting molecules into local patterns characterized by strong bonds and nearly tetrahedral angles and more compact arrangements characterized by more strain and bond breakage.

According to a model proposed by Wiggins (1990), two types of water structure can be distinguished: high-density water and low-density water. In the dense water, the bent, relatively weak hydrogen bonds predominate over straight stronger ones. Low-density water has many ice-like straight hydrogen bonds. Although hydrogen bonding is still continuous through the liquid, the weakness of the bonds allows the structure to be disrupted by thermal energy extremely rapidly. High-density water is extremely reactive and more liquid, whereas low-density water is inert and is more viscous. A continuous spectrum of water structures between these two extremes could be imagined. The strength of water–water hydrogen bonding, which is the source of water density and reactivity, has great functional significance; this explains solvent water's properties and its role in many biological events.

A common feature of all theories is that a definite structure of liquid water is due to the hydrogen bonding between molecules and that the structure is in the dynamic state as the hydrogen bonds break and reform with high frequency.

WATER AS A SOLVENT

Many molecular parameters, such as ionization, molecular and electronic structure, size, and stereochemistry, will all influence the basic interaction between a solute and a solvent. The addition of any substance to water results in altered properties for this substance and for water itself. Solutes cause a change in water properties because the hydrate envelopes that are formed around dissolved molecules are more organized and therefore more stable than the flickering clusters of free water. The properties of solutions that depend on solute and its concentration are different from those of pure water. The differences can be seen in such phenomena as the freezing point depression, boiling point elevation, and increased osmotic pressure of solutions.

The polar nature of the water molecule and the ability to form hydrogen bonds determine its properties as a solvent. Water is a good solvent for charged or polar compounds and a relatively poor solvent for hydrocarbons. Hydrophilic compounds interact strongly with water by an ion–dipole or dipole–dipole mechanism causing changes in water structure and mobility and in the structure and reactivity of the solutes. The interaction of water with various solutes is referred to as hydration. The extent and tenacity of hydration depends on a number of factors, including the nature of the solute, salt composition of the medium, pH, and temperature.

Water dissolves dissociable solutes readily, because the polar water molecules orient themselves around ions and partially neutralize ionic charges. As a result of that, the positive and negative ions can exist as separate entities in a dilute aqueous solution without forming ion pairs. Sodium chloride is shown as an example where the electrostatic attraction of the Na^+ and Cl^- is overcome by the attraction of the Na^+ with the negative charge on the oxygens and the Cl^- with the positive charge on the hydrogen ions (Figure 2.8). The number of weak charge–charge interactions between water and the Na^+ and Cl^- ions is sufficient to separate the two charged ions from the crystal lattice (Lehninger et al., 1993).

To acquire their stabilizing hydration shell, ions must compete with water molecules, which need to make as many hydrogen bonds with one another as possible. The normal structure of pure water is disrupted in

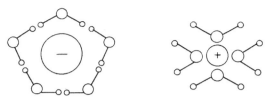

Figure 2.8 Hydration shell around Na⁺ and Cl⁻.

solution of dissociable solutes. The ability of a given ion to alter the net structure of water is dependent on the strength of its electric field. Among ions of a given charge type, e.g., Na^+ and K^+ or Mg^{+2} and Ca^{+2}, the smaller ions are more strongly hydrated than the larger ions in which the charge is dispersed over a greater surface area. Most cations, except the largest ones, have a primary hydration sphere containing four to six molecules of water. Other water molecules, more distant from the ion, are held in a looser secondary sphere. The electrochemical transfer experiments indicate a total of sixteen molecules of water around Na^+ and about ten around K^+. The bound water is less mobile and more dense than HOH molecules in the bulk water. At some distance, the bonding arrangements melt into dynamic configuration of pure water.

Water is especially effective in screening the electrostatic interaction between dissolved ions, because by Coulomb's law, the force (F) between two charges q^+ and q^-, separated by a distance r, is given as:

$$F = \frac{q^+ \cdot q^-}{\epsilon r^2} \tag{2.1}$$

where ϵ is the dielectric constant of the medium. For vacuum, $\epsilon = 1$ Debye unit whereas bulk water $\epsilon = 80$; it implies that the energies associated with electrostatic interactions in aqueous media are approximately 100 times smaller than the energies of covalent association but increase considerably in the interior of a protein molecule (Nossel and Lecar, 1991).

In thermodynamic terms, the free energy change, ΔG, must have a negative value for a process to occur spontaneously.

$$\Delta G = \Delta H - T\Delta S \tag{2.2}$$

where ΔG represents the driving force, ΔH (the enthalpy change) is the energy from making and breaking bonds, and ΔS (the entropy change) is the increase in randomness.

Solubilization of a salt occurs with a favorable change in free energy. As salt such as NaCl dissolves, the Na^+ and Cl^- ions leaving the crystal lattice acquire greater freedom of motion. The entropy of the system increases; where ΔH has a small positive value and $T\Delta S$ is large and positive, ΔG is negative.

Water in the multilayer environment of ions is believed to exist in a structurally disrupted state because of conflicting structural influences of the innermost vicinal water and the outermost bulk-phase water. In concentrated salt solutions, the bulk-phase water would be eliminated, and water structure common in the vicinity of ions would predominate. Small or multivalent ions such as Li^+, Na^+, H_3O^+, Ca^{+2}, Mg^{+2}, F^-, SO_4^{-2}, PO_4^{-3}, which have strong electric fields, are classified as water structure formers because solutions containing these ions are less fluid than pure water. Ions that are large and monovalent, most of the negatively charged ions and large positive ions such as K^+, Rb^+, Cs^+, NH_4^+, Cl^-, Br^-, I^-, NO_3^-, ClO_4^-, CNS^-, disrupt the normal structure of water; they are structure breakers. Solutions containing these ions are more fluid than pure water (Fennema, 1985).

Ions influence all kinds of water solute interactions through their varying abilities to hydrate and to alter water structure and its dielectric constant. The conformation of macromolecules and stability of colloids are greatly affected by the kinds and concentrations of ions present in the medium.

Water is a good solvent for most biomolecules, which are generally charged or polar compounds. Solubilization of compounds with functional groups such as ionized carboxylic acids (COO^-), protonated amines (NH_3^+), phosphate esters, or anhydrides is also a result of hydration and charge screening.

Uncharged but polar compounds possessing hydrogen bonding capabilities are also readily dissolved in water due to the formation of hydrogen bonds with water molecules. Every group that is capable of forming a hydrogen bond to another organic group is also able to form hydrogen bonds of similar strength with water. Hydrogen bonding of water occurs with neutral compounds containing hydroxyl, amino, carbonyl, amide, or imine groups. Saccharides dissolve readily in water due to the formation of many hydrogen bonds between the hydroxyl groups or carbonyl oxygen of the saccharide and water molecules. Water-solute hydrogen bonds are weaker than ion–water interactions. Hydrogen bonding between water and polar solutes also causes some ordering of water molecules, but the effect is less significant than with ionic or nonpolar solutes.

The introduction into water of hydrophobic substances such as hydrocarbons, rare gases, and the apolar groups of fatty acids, amino acids, or proteins is thermodynamically unfavorable because of the decrease in entropy. The decrease in entropy arises from the increase in water to water hydrogen bonding adjacent to apolar entities. Water molecules in the immediate vicinity of a nonpolar solute are constrained in their possible orientations, resulting in a shell of highly ordered water molecules around each nonpolar solute molecule [Figure 2.9(a)]. The number of water molecules in the highly ordered shell is proportional to the surface area of the hydrophobic solute. In the case of dissolved hydrocarbons, the enthalpy of formation of the new hydrogen bonds often almost exactly balances the enthalpy of creation in water, a cavity of the right size to accommodate the hydrophobic molecule. However, the restriction of the water mobility results in a very large decrease in the entropy. According to $\Delta G = \Delta H - T\Delta S$, if ΔH is almost zero and ΔS is negative, ΔG is positive.

To minimize the contact with water, hydrophobic groups tend to aggregate; this process is known as hydrophobic interaction [Figure 2.9(b)]. The existence of hydrophobic substances barely soluble in water, but readily soluble in many nonpolar solvents, and their tendency to aggregate in aqueous media have been known for a long time. However the origin of this hydrophobic effect is still somewhat controversial. The plausible explanation is that hydrophobic molecules disturb the hydrogen bonded state

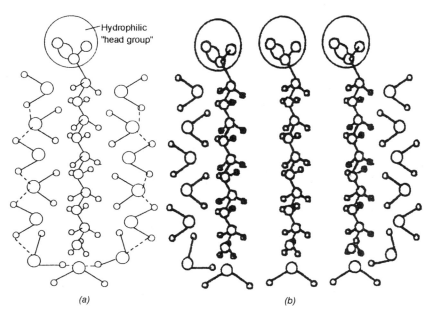

(a)   (b)

Figure 2.9 Cage-like water structure around the (a) hydrophobic alkyl chains and (b) hydrophobic interactions.

of water without having any compensatory ordering effects. Apolar molecules are water structure formers: water molecules cannot use all four possible hydrogen bonds when in contact with hydrophobic, water-hating molecules. This restriction results in a loss of entropy, a gain in density, and increased organization of bulk water.

Amphipathic molecules, that is compounds containing both polar or charged groups and apolar regions, disperse in water if the attraction of the polar groups for water can overcome possible hydrophobic interactions of the apolar portions of the molecules. Many biomolecules are amphipathics: proteins, phospholipids, sterols, certain vitamins, and pigments, have polar and nonpolar regions. When amphipathic compounds are in contact with water, the two regions of the solute molecule experience conflicting tendencies: the polar or charged hydrophilic regions interact favorably with water and tend to dissolve, but the nonpolar, hydrophobic regions tend to avoid contact with water. The nonpolar regions of the molecules cluster together to present the smallest hydrophobic area to the aqueous medium, and the polar regions are arranged to maximize their interactions with the aqueous solvent. Many amphipathic compounds are able to form in aqueous media stable structures, containing hundreds to thousands of molecules called micelles. The forces that held the nonpolar regions of the molecules together are due to hydrophobic interactions (Lehninger et al., 1993; Nossal and Lecar, 1991).

The hydrophobic effect is a driving force in formation of clathrate hydrates and self-assembly of lipid bilayers. Hydrophobic interactions between lipids and proteins are the most important determinants of biological membrane structure. The three-dimensional folding pattern of proteins is also determined by hydrophobic interactions between nonpolar side chains of amino acid residues.

Water can be expected to form locally ordered structures at the surface of water-soluble, as well as water-insoluble, macromolecules and at the boundaries of the cellular organelles. Water at the interfaces and surfaces may have quite different properties from those that it has in bulk phase. Biomacromolecules generally have many ionized and polar groups on their surfaces and tend to align near polar water molecules. This ordering effect exerted by macromolecular surface extends quite far into the surrounding medium. Various physical techniques (i.e., nuclear magnetic resonance) and chemical probes (exchange of H by D) indicate that there is a layer of water that is bound to protein molecules, phosopholipid bilayers, or nucleic acids.

Measurements of the diffusion coefficients of globular protein molecules in solution yield values for molecular size that are greater than the corresponding radii determined by X-ray crystallography. The apparent hydrodynamic radius can be calculated from the Stokes-Einstein relation:

$$D = k_B \cdot \frac{T}{6\pi\eta a_H} \qquad (2.3)$$

where D is the diffusion coefficient, k_B is the Boltzmann constant, T is the temperature, η is the solution viscosity, and a_H is the molecule radius (Nossal and Lecar, 1991).

Similarly, studies utilizing nuclear magnetic resonance techniques show that there is a species of associated water that has a different character than does the water in the bulk phase. By these and other methods, it was found that, for a wide range of protein molecules, approximately 0.25–0.45 g of H_2O is associated with each gram of protein.

It has been traditional to refer to the water that is associated with proteins and other macromolecules as bound water. However, to designate such water as bound can be misleading, because for the most part, the water molecules probably are only transiently associated, and at least a portion of the associated water has to be constantly rearranged due to the thermal perturbations of weak hydrogen bonds.

The hydration forces can stabilize macromolecular association or prevent macromolecular interactions with a strength that depends on the surface characteristic of the molecules and ionic composition of the medium. Biophysical processes involving membrane transport also are influenced by hydration. The size of the hydration shell surrounding small ions and the presence of water in the cavities of ionic channels or in the defects between membrane lipids strongly affect the rates at which the ions cross a cell membrane.

The interaction between a solute and a solid phase is also influenced by water. Hydration shells or icebergs associated with one or the other phase are destroyed or created in this interaction and often contribute to conformational changes in macromolecular structures and ultimately to changes in biological and functional properties important in food processing.

INTRACELLULAR WATER

The idea that intracellular water exhibits peculiar properties different from those of bulk water has been around for a long time. The uniqueness of the cytoplasmic water was deduced from

- the observations that cells may be cooled far below the freezing point of a salt solution iso-osmotic with that of the cytoplasm
- properties of the cytoplasm, which in some condition, should bind water forming a gel
- osmotic experiments in which it has often been observed that part of the cell water is not available as a solvent; this water has been

described as osmotically inactive, bound water, or
compartmentized water

According to a recent view, three different kinds of intracellular water
can be distinguished: a percentage of total cell water appears in the form
of usual liquid water. A relevant part is made up of water molecules that
are bound to different sites of macromolecules in the form of hydration
water, while a sizeable amount, although not fixed to any definite molecular site, is strongly affected by macromolecular fields. This kind of water
has been termed "vicinal water." Most of the vicinal water surrounds the elements of the cell cytoskeleton. Vicinal water has been extensively investigated, and it has been found that some of its properties are different
from those of normal water. It does not have a unique freezing temperature, but an interval ranging from -70 to $-50°C$; it is a very bad solvent
for electrolytes, but nonelectrolytes have the same solubility properties in
it as in usual water; its viscosity is enhanced, and its NMR response is
anomalous (Giudice et al., 1986).

The distribution of various types of water inside the living cells is a
question that cannot be answered today, especially because in many cells,
marked changes have been noted in the state of intracellular water as a
result of biological activity. The possibility that water in living cells may
differ structurally from bulk water has incited a search for parameters of
cell water, which deviate numerically from those of bulk water.

The diffusion coefficient for water in cytoplasm of various cells has been
determined with a satisfactory precision. It has been found that the movement of water molecules inside living cells is not very different and is
reduced by a factor between two and six, as compared to the self-diffusion
coefficient for pure water. According to Mild and Løvtrup (1985), the most
likely explanation of the observed values is that part of the cytoplasmic
water, the vicinal water, is structurally changed to the extent that its rate of
motion is significantly reduced as compared to bulk phase.

WATER TRANSPORT

When cells are exposed to hyper- or hypo-osmotic solutions, they immediately lose or gain water, respectively. Even in an isotonic medium, a
continuous exchange of water occurs between living cells and their surroundings. Most cells are so small and their membrane so leaky that the
exchange of water measured with isotopic water reaches equilibrium in a
few milliseconds.

The degree of water permeability differs considerably between tissues
and cell types. Mammalian red blood cells and renal proximal tubules are

extremely permeable to water molecules. Transmembrane water movements are involved in diverse physiological secretion processes.

How water passes through cells has only begun to come clear in the last five years. Water permeates living membranes through both the lipid bilayer and through specific water transport proteins. In both cases, water flow is passive and directed by osmosis. Water transport in living cells is therefore under control of ATP and ion pumps.

The most general water transport mechanism is diffusion through lipid bilayers. The diffusion through a lipid bilayer depends on the lipid structure and presence of sterols and occurs only when lipids are in the liquid state. It is suggested that the lateral diffusion of the lipid molecules and the water diffusion through membrane is a single process (Haines, 1994).

A small amount of water is transported through certain membrane transport proteins such as glucose transporter or the anion channel of erythrocytes. The major volume of water passes through specific water transport proteins. The first isolated water-transporting protein was the channel-forming integral protein from red blood cells. The identification of this protein has led to the recognition of a family of related water-selective channels, the aquaporins, that are found in animals, plants, and microbial organisms. Water flow through protein channel is controlled by the number of copies of the proteins in the membrane. In red blood cells, there are 200,000 copies per cell, and in apical brush border cells of renal tubules it constitutes 4% of the total protein (Engel et al., 1994).

WATER IN FOOD

PROPERTIES

Water, with a density of 1,000 kg/m³, is denser than the oil components of foods; oils and fats typically have densities in the range of 850–950 kg/m³. Glycerols and sugar solutions are denser than water.

Unlike solid phases of most other liquids, ice is less dense than liquid water; ice has a lower thermal conductivity than water. These properties have an effect on the freezing of foods that are predominantly water based; the formation of an ice layer on the surface of the liquids and the outside of solids has the effect of slowing down the freezing rate.

Because a molecule of water vapor is lighter (molecular weight = 18) than that of dry air (molecular weight = about 29), moist air is lighter than dry air at the same temperature. This is somewhat unexpected in that the popular conception is that humid air (which contains more water) is heavier that dry air.

At room temperature, water has the highest specific heat of any inorganic or organic compound, with the sole exception of ammonia. It is interesting to speculate why the most commonly occurring substance on this planet should have one of the highest specific heats. One of the consequences of this peculiarity in the food industry is that heating and cooling operations for essentially water-based foods are more energy demanding. To heat a kilogram of water from 20°C to 50°C requires about 125 kJ of energy, whereas heating the same mass of vegetable oil requires only 44 kJ.

WATER–FOOD SYSTEMS

A sponge holds most of its water as liquid held in the inner structure of the sponge. Most of the water can be wrung out of the sponge, leaving a matrix of air and damp fibers. Within the sponge fibers, the residual water is more tenuously held—absorbed within the fiber of the sponge. If the sponge is left to dry in the sun, this adsorbed water will evaporate, leaving only a small proportion of water bound chemically to the salts and to the cellulose of the sponge fibers. As with the familiar example of water in a sponge, water is held in food by various physical and chemical mechanisms.

It is a convenient oversimplification to distinguish between free and bound water. The definition of bound water in such a classification poses problems. Fennema (1985) reports seven different definitions of bound water. Some of these definitions are based on the freezability of the bound component, and others rely on its availability as a solvent. He prefers a definition in which bound water is "that which exists in the vicinity of solutes and other nonaqueous constituents, exhibits reduced molecular activity and other significantly altered properties as compared with bulk water in the same system, and does not freeze at −40°C." Table 2.1 illustrates Fennema's classification for water states in foods.

MOISTURE CONTENT

Moisture content can be measured simply by weighing a sample and then oven drying it, usually at 105°C overnight with the difference in mass being the moisture in the original sample. However, much confusion is caused by reporting the moisture content simply as a percentage without specifying the basis of the calculation. It should be made clear whether the moisture content is calculated on a wet basis (moisture content divided by original mass) or on a dry basis (moisture content divided by the "bone dry" or "oven dry" mass). Even the term "bone dry" mass can cause confusion amongst non-English speakers; it was once misinterpreted as the

TABLE 2.1. Classification of Water States in Foods (after Fennema, 1985).

Class of Water	Description	Proportion of Typical 90% (wet basis) Moisture Content Food
Constitutional	An integral part of nonaqueous constituent	Less than 0.03%
Vicinal	Bound water that strongly acts with specific hydrophilic sites of non-aqueous constituents to form a monolayer coverage; water–ion and water–dipole bonds	0.1 to 0.9%
Multilayer	Bound water that forms several additional layers around hydrophylic groups; water–water and water–solute hydrogen bonds	1 to 5%
Free	Flow is unimpeded; properties close to dilute salt solutions; water–water bonds predominate	5% to about 96%
Entrapped	Free water held within matrix or gel, which impedes flow	5% to about 96%

"mass of the dry bones"! In foods containing significant quantities of fat or salt for example, moisture content may be calculated as the mass of water in a sample divided by the dry solids which are not salt or fat—in which case the moisture content should be reported as "salt free, fat free, dry basis."

SORPTION ISOTHERMS AND WATER ACTIVITY

Since 1929, it has been recognized that the chemical and microbial stability and, hence, the shelf life of foods are not directly related to its moisture content, but to a property called water activity (a_w) (Tomkins, 1929). Essentially a_w is a measure of the degree to which water is bound within the food and, hence, is unavailable for further chemical or microbial activity.

The definition of a_w is in terms of one of the ways it can be measured. It is defined as the ratio of the partial pressure of water vapor in or around the food to that of pure water at the same temperature. Relative humidity of moist air is defined in the same way, except that, by convention, relative humidity is reported as a percentage, whereas a_w is expressed as a fraction. Thus, if a sample of meat sausage is sealed within an airtight container, the humidity of the air in the headspace will rise and eventually equilibrate to a relative humidity of about 83%, which means that the a_w of the meat sausage is 0.83.

The relationship between a_w and moisture content for most foods at a particular temperature is a sigmoidal-shaped curve called the sorption isotherm (Figure 2.10). The phrase equilibrium moisture content curve is also used. Sorption isotherms at different temperatures can be calculated using the Clausius-Clapeyron equation from classical thermodynamics, namely:

$$\frac{d(\ln a_w)}{d(1/T)} = \frac{\Delta H}{R} \qquad (2.4)$$

where T is the absolute temperature, ΔH is the heat of sorption and R is the gas constant.

A complication arises from one of the methods of measuring sorption isotherms for a food. A food that has previously been dried and then is re-hydrated will have a different sorption isotherm (adsorption isotherm) from that which is in the process of drying (desorption isotherm). This difference is due to a change in water-binding capacity in foods that have been previously dried.

Many mathematical descriptors for sorption isotherms have been proposed. One of the more famous is that of Brunauer, Emmett, and Teller (1938), the B.E.T. isotherm, which is based on the concept of a measurable amount of monomolecular layer (vicinal) water for a particular food. Wolf et al. (1985) compiled 2,201 references to sorption isotherm data for foods.

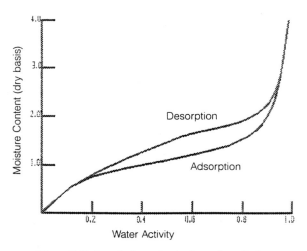

Figure 2.10 A typical sorption isotherm for a food.

TABLE 2.2. Sorption Isotherm Data for Cod and Corn
(from Iglesias et al., 1975).

Product	Specifications	Temp. (°C)	X_m	r	a'	Reference
Cod	Adsorption	30	7.68	1.2398	1.3490	Jason (1958)
Corn	Desorption	4.5	8.30	2.2345	1.9748	Chen & Clayton (1971)
	Desorption	15.5	7.68	2.4862	2.0949	Chen & Clayton (1971)
	Desorption	30	7.30	2.5663	1.7950	Chen & Clayton (1971)
	Desorption	38	6.35	2.3711	1.8618	Chen & Clayton (1971)
	Desorption	50	6.89	2.1203	1.5936	Chen & Clayton (1971)
	Desorption	60	5.11	2.2185	1.7430	Chen & Clayton (1971)

An example of the type, detail, and accuracy of sorption isotherm data available in the literature is presented in Table 2.2.

Iglesias et al. (1975) propose the following three-parameter equation to fit sorption isotherm data for a range of foods:

$$a_w = \exp(-a'\theta^r) \tag{2.5}$$

where a' and r are the parameters as listed in Table 2.2 and $\theta = X/X_m$. X is the equilibrium moisture content and X_m, in units of g/100 g dry basis, is the B.E.T. monomolecular moisture content for the particular food as listed in Table 2.2. However, there are nearly as many equations to sorption isotherms as there are researchers in this field.

WATER ACTIVITY AND SHELF LIFE OF FOODS

Many of the chemical and biological processes that cause deterioration of foods, and ultimately spoilage, are water-dependent. Microbial growth is directly linked to a_w. No microbes can multiply at an a_w below 0.6. Dehydration is arguably the oldest form of food preservation; the sun drying of meat and fish has been traced to the beginning of recorded history. Drying relies on removing water, thus making it unavailable for microbial growth.

Salting or curing has the same effect. A saturated solution of common salt has an a_w close to 0.75. Thus, by adding sufficient salt to foods, the a_w

can be lowered to a level where most pathogenic bacteria are inactivated, but the moisture content remains high.

Intermediate moisture content foods (IMF) such as pet food and continental sausages rely on fats and water-binding humectants such as glycerol to lower a_w. Fat, being essentially hydrophobic, does not bind water but acts as a filler for IMF to increase the volume of the product.

The effect of several humectants is for each to sequester an amount of water independently of the other humectants that may be present in the food. Each thus lowers the a_w of the system according to the Ross (1975) equation:

$$a_{wn} = a_{w0} \cdot a_{w1} \cdot a_{w2} \cdot a_{w3} \cdot \text{etc.} \qquad (2.6)$$

where a_{wn} is the water activity of the complex food system and a_{w0}, etc. are the water activities associated with each component of the system.

For example, a_w of a food with a moisture content of 77% (wet basis) and a salt content of 3% (wet basis) can be calculated as follows: 100 g of the food comprises 77 g of water, 20 g of "bone" dry matter, and 3 g of salt. The contribution to a_w due to the salt can be calculated (according to Raoult's law of dilute solutions) and the molecular weights of water (18) and salt (58.5) as:

$$a_{w1} = (77 \times 18)/(77 \times 18 + 3 \times 58.5) = 0.89$$

The a_w for the salt-free solid matter of the food is found from its sorption isotherm at that moisture content, $a_{w0} = 0.90$, say. Thus, the a_w of the salted food is

$$a_{wn} = a_{w0} \cdot a_{w1} = 0.9 \times 0.89 = 0.8$$

None of the dangerous pathogenic bacteria associated with food such as *Clostridium* or *Vibrio* spp., which cause botulism and cholera, respectively, can multiply at a_w values below about 0.9. Thus, drying, or providing sufficient water-binding humectants, is an effective method of preventing the growth of food poisoning bacteria.

Only osmophilic yeasts and some molds can grow at a_w in the range 0.6 to 0.65. Thus, by reducing the a_w below this value, foods are microbially stable. That is unless the packaging is such that the food becomes locally rewet, in which case local spoilage can occur, for example, when condensation occurs within a hermetically sealed package subject to rapid cooling.

There are various chemical reactions that proceed and may be accelerated at low values of a_w. Maillard reactions leading to lysine loss and brown color development peak at a_w values around 0.5 to 0.8. Nonenzymatic lipid oxidation increases rapidly below $a_w = 0.4$. Enzymic hydrolysis decreases with a_w down to $a_w = 0.3$ and is then negligible.

WATER SUPPLY, QUALITY, AND DISPOSAL

WATER SUPPLY

Just as water is an integral part of any food, the supply, quality, and disposal of water is of prime consideration in the establishment and operation of all food processing. Potable (drinkable) water may be required for addition to the product and will certainly be required for clean-up. Nonpotable water may be used for heat exchangers and cooling towers. Boiler feed water must be conditioned within close limits of pH and hardness. Brennan et al. (1990), in their book *Food Engineering Operations,* list four types of water used in the food and beverage industries:

(1) General purpose water

(2) Process water

(3) Cooling water

(4) Boiler feed water

The siting and consequent viability of a food processing plant may well depend on a guaranteed, regular supply of suitable quality water and an environmentally acceptable method of disposal. Developed countries now have strict regulations for the emission of wastewater. Developing countries are becoming increasingly aware of the problems of wastewater disposal. In a recent symposium in Indonesia, a fish drying processor was asked what his main technical problems were. He nominated water pollution—not for reasons of meeting environmental control regulations, but because the fish farmers further down the river were complaining about his wastewater.

WATER QUALITY

There are a number of international standards for potable water quality in existence. The World Health Organization (WHO) has a standard for potable water quality as part of the *Codex Alimentarius*. The standard detailed in Table 2.3 is from the United States Environmental Protection Agency. There is also a large EC directive relating to the quality of water intended for human consumption (80/778/EEC).

TABLE 2.3. **Primary Maximum Contaminant Levels in Potable Water (mg/L unless specified).**

Arsenic 0.05	Barium 1
Cadmium 0.010	Chromium 0.05
Lead 0.05	Mercury 0.002
Nitrate (as N) 10	Selenium 0.01
Silver 0.05	Fluoride 4.0
Endrin 0.0002	Lindane 0.004
Methoxychlor 0.1	Toxaphene 0.005
2.4 D 0.1	2,4,5 TP Silvex 0.01
Total trihalomethanes 0.10	Trichloroethylene 0.005
Carbon tetrachloride 0.005	1,2-dichloroethane 0.005
Vinyl chloride 0.002	Benzene 0.005
para-Dichlorobenzene 0.075	1,1-Dichloroethylene 0.007
1,1,1-trichloroethane 0.2	Radium 226 land 228, combined 5 pC/1
Gross alpha particle activity 15 pC/1	Gross beta particle activity 4 millirem/year
Turbidity 1 tu up to 5 tu average	
Coliform bacteria 1 per 100 ml, monthly	

WATER TREATMENT

In most cases, water will require some treatment to assure it is of the required standard to meet food hygiene requirements and not constitute a public health hazard. Surface water from rain runoff into rivers or impoundments is likely to contain atmospheric solutes, minerals from the ground, organic matter from vegetation, microbial contamination from birds and wild and domestic animals, and human waste. Water from underground aquifers will have much of the surface contamination filtered out, but it is likely to be high in dissolved mineral content.

Water treatment to bring the quality to within the required standard may involve screening, sedimentation, coagulation and flocculation, filtration, and other physical or chemical treatments to remove microorganisms, organic matter, or dissolved minerals.

Metal screens are used to remove particles larger than about 1 mm in size. Settling ponds remove smaller particles. Insoluble, suspended matter is usually removed by sand filters. Coagulating and flocculating agents act to bind smaller particles into clumps that then settle or can be screened or filtered.

Microorganisms can be inactivated by heat, chemical disinfection, UV radiation, or ultrasonic treatment. Most town water supplies are chlorinated or have ozone added for chemical disinfection.

Treatment to remove dissolved mineral matter is more complex. Dissolved bicarbonates of calcium, magnesium, sodium, and potassium cause alkalinity; soluble calcium and magnesium salts cause hardness. Alkalinity and hardness may need to be adjusted for some food processing operations. For example, the formation of a "head" on beer is critically dependent on water hardness. Excessively hard water may cause discoloration and toughening of certain foods. On the other hand, hardness may be required to prevent excessive foaming in clean-up operations.

Iron and manganese salts may be present in water supplies forming organic slimes, which tend to clog pipes. Aeration, filtering, and settling are effective for the removal of iron bicarbonates. Insoluble oxides of manganese are formed through chlorination.

Excessive amounts of dissolved gases, carbon dioxide, oxygen, nitrogen, and hydrogen sulfide cause problems in boiler feed water, corrosion, and bacterial formation. Treatment is by boiling and venting off the noncondensable gases or by chemical dosing.

Small amounts of hydrocarbons such as kerosene and diesel cause tainting in foods. Separating fuels from processing areas, personal hygiene, cleaning stations around food processing operations, and good housekeeping can prevent this problem.

WASTEWATER TREATMENT

The ultimate aim of any food processing operation is to have an environmentally neutral impact. Reuse, recycle, and sustainability are today's catchwords. For a food operation to be truly environmentally sustainable, it should recycle all water not incorporated in the product or vented to the atmosphere. The reality is that it is currently considered uneconomic to recycle wastewater from food processing operations. Current practice is to treat wastewater to limit its effect on receiving waters. Treatment of wastewater mirrors the water treatment methods described above, which is a combination of physical, chemical, and biological treatments.

The main problem with food waste is its high biochemical oxygen demand, or BOD. Food wastes contain large quantities of organic matter, which break down naturally by oxidation; however, this oxygen demand is at the expense of other natural biochemical processes in waterways that become oxygen depleted and lifeless if the BOD is too high. BOD is defined as the quantity of oxygen (in units of mg/L) required for a microorganism to oxidize the waste at a particular temperature (20°C) in 5 days. Food wastes can range in BOD from 500–4,000 mg/L BOD, which is higher than for domestic sewage (200–400 mg/L).

Wastewater from food processing operations usually contains significant solid matter that can be removed by screens or by sedimentation. Fats and oils can be skimmed from the surface of settling tanks; heavier suspended matter is removed as sludge, which can then be dewatered, dried, and used as animal feed, fertilizer, or fuel. An alternative method for the removal of oils and fats is by aeration, in which air bubbles blown from the bottom of a settling tank carry fine solids and grease to the surface.

Chemical treatments of wastewater are much the same as described above for process water treatment.

BOD can be effectively reduced by biological treatment. Both aerobic and anaerobic fermentation of the organic material is used. Depending on the scale of the operation, bioreactors range in size from 7.5 m diameter and 2–3 m in depth to lagoons 1–2 m deep covering several hectares. However, for long-term sustainable operation, there must be provision for sludge removal. Where area for treatment is not a limitation and there is sufficient isolation for smell not to be a deterrent, wastewater is sprayed directly on the ground, where it breaks down under the action of sunlight and in-ground bacteria.

Sewage and wastewater can also be treated anaerobically. Closed reactors facilitate odor control, although anaerobic lagoons are also used. Such lagoons are deeper than for aerobic types, with grease allowed to accumulate on the surface to control odor emission. Methane produced as an end product of the biochemical pathway can be used for heating the reactors in cold weather. A problem with the anaerobic digestion process is its sensitivity to pH and temperature variation and the susceptibility of the active microorganisms to chemical disinfectants.

REFERENCES

Brunauer, S., Emmet, P. H. and Teller, E. 1938. "Adsorption of gases in multilayers." *Journal of the American Chemical Society,* 60(2):309–319.

Brennan, J. G., Butters, J. R., Cowell, N. D. and Lilley, A. E. V. 1990. *Food Engineering Operations.* 3rd Edition. Elsevier Applied Science, London and New York, p. 523.

Chen, C. S. and Clayton, J. T. 1971. "The effect of temperature on sorption isotherms of biological materials." *Trans. A.S.A.E.* 14(5):927–929.

Engel, A., Waltz, Th. and Agre, P. 1994. "The Aquaporin Family of Membrane Water Channels," *Current Opinion in Structural Biology,* 4:545–553.

Fennema, O. R. 1985. *Food Chemistry.* 2nd Edition. Marcel Dekker Inc., New York and Basel.

Giudice E., Doglia, S., Milani, M. and Vitiello, G. 1986. "Water in Biological Systems," in *Modern Bioelectrochemistry,* F. Gutmann and H. Keyzer, eds. Plenum Press, New York and London, p. 282.

Haines, Th. H. 1994. "Water transport across biological membranes." *FEBS Letters,* 346:115–122.

Iglesias, H. A., Chirife, J. and Lombardi, J. L. 1975. "An equation for correlating equilibrium moisture content in foods." *Journal of Food Technology,* 10:289–297.

Jason, A. C. 1958. "A study of evaporation and diffusion processes in the drying of fish muscle." In: *Fundamental Aspects of the Dehydration of Foodstuffs,* Soc. Chem. Ind. pp. 103–135.

Lehninger, A. L., Nelson, D. L. and Cox, M. M. 1993. *Principles of Biochemistry.* Worth Publishers, Inc. pp. 81–88.

Mild, K. H. and Lovtrup, S. 1985. "Movement and structure of water in animal cells. Ideas and experiments," *Biochim. Biophys. Acta,* 822:155–167.

Nossal, R. and Lecar, H., eds. 1991. *Molecular and Cell Biophysics.* Addison-Wesley Publishing Company, pp. 9–40.

Ross, K. D. 1975. "Estimation of water activity in intermediate moisture foods." *Food Technology,* 29(3):26–34.

Stillinger, F. H. 1980. "Water revisited," *Science,* 209:451–457.

Tomkins, R. G. 1929. "Studies of the growth of moulds. 1." *Proceedings of the Royal Society B,* 105:375–401.

USDPA. 1987. *The Safe Drinking Water Act.* United States Departmental Protection Agency, Program Summary, October 1987.

Wiggins, Ph. M. 1990. "Role of water in some biological processes," *Microbiol. Rev.,* 54:432–449.

Wolf, W., Spiess, W. E. L. and Jung, G. 1985. *Sorption Isotherms and Water Activity of Food Materials.* Science and Technology Publishers, Hornchurch, Essex, England.

CHAPTER 3

Mineral Components

MICHAŁ NABRZYSKI

THE CONTENTS AND ROLE OF MINERALS IN FOODS

MINERALS represent from 0.2% to 0.3% of the total intake of all nutrients in the diet. They are so potent and so important that, without them, the organism wouldn't be able to utilize all the other 99.7% of the food. The main mass of these minerals constitutes the macroelements, and the trace elements constitute only a hundredth of the percent of the total mass of daily eaten nutrients. The foods that are good sources of some minerals are given in Table 3.1.

Dietary minerals are necessary for maintenance of normal cellular metabolism and tissue function. These nutrients participate in a multitude of biochemical and physiological processes important for health. Because of their broad biochemical activity, many of these compounds are intentionally used as functional agents in a variety of foods. On the other hand, some cations may also induce a diversity of undesirable effects that influence the nutritional quality of foods.

Minerals play an important role in plant life. They function as catalysts of biochemical reactions, are responsible for changes in the state of cellular colloids, directly affect the cell metabolism, and are involved in changes in protoplasm turgor and permeability. They often become the centers of electrical and radioactive phenomena in living organisms.

Minerals are usually grouped in two categories: the macroelements required in our diets in amounts greater than 100 mg and the microelements required in milligram quantities or less per day. The macroelements include calcium, magnesium, phosphorus, sodium, potassium, sulphur, and chlorine. The microelements are comprised of iron, zinc, copper, manganese, iodine, cobalt, nickel, molybdenum, chromium, fluorine,

TABLE 3.1. **The Contents of Selected Minerals in Some Foods.**

Mineral	Food	Amount (mg/100 g)	Reference*
Calcium	Swiss cheese—low sodium	960	2
	Sardines in tomato sauce	437**	
	Cod in tomato sauce	335**	
	Yogurt natural	189**	
	Milk	120**	
	Orange	42	11
	Carrot	41	11
	Tuna in own sauce	25**	
	Potato	10	11
Potassium	Wheat seeds	502	11
	Porcine liver	350	11
	Beef	342	11
	Oat, flaked	335	11
	Carrot	290	11
	Pork	260	11
	Orange	177	11
	Milk	157	11
	Wheat meal (550)	126	11
	Cheese (45% fat)	107	11
Magnesium	Sardine in tomato sauce	27**	
	Tuna in own sauce	24**	
	Yogurt natural	12**	
	Milk	9**	
Iron	Porcine liver	22	11
	Egg (yolk)	7.2	11
	Oat, flaked	4.6	11
	Wheat seeds	3.3	11
	Pork	2.3	11
	Beef	2.6	11
	Wheat meal (550)	1.1	11
	Egg (white)	0.2	11

(continued)

TABLE 3.1. (continued).

Mineral	Food	Amount (mg/100 g)	Reference*
Copper	Oysters	6; 17	4; 9
	Liver, calf	7	9
	Liver, beef	3	9
	Wheat germ	0.9; 2	5; 9
	Sunflower seeds	2	9
	Tuna	0.5	9
	Salmon	0.2	9
	Ham	0.03; 0.08	9; 5
Chromium	Spices	>0.1–0.5	10
	Cacao	0.2	10
	Paprika, pepper, curry	~0.05	10
	Hawthorn	0.025	10
	Cheese drowned	>0.01	10
	Whole meal bread	~0.02	10
	Beef	<0.004	10
	Kidney, liver	<0.0015	10
Fluoride	Black teas	3–34	6
	Fish canned	0.09–0.8	8
	Shellfish	0.03–0.15	8
Iodide	Marine fish, oyster, shrimp, lobster	0.02–0.1	1
	Milk powder	0.06	7
Selenium	Tortilla chips	1.0	2
	Potato chips	0.97	2
	Pork kidney, braised	0.21	2
	Tuna, canned	0.12	2
	Salmon, canned	0.08	2
	Milk	0.002	2

*Data taken from: 1) Causeret, 1962; 2) Feltman, 1990; 3) Gajek et al., 1987; 4) Lopez et al., 1983; 5) Marzec et al., 1992; 6) Nabrzyski and Gajewska, 1995; 7) Paslawska and Nabrzyski, 1975; 8) WHO Environmental Health Criteria, 1984; 9) Williams, 1982; 10) Wilpinger et al., 1995; 11) Wojnowski, 1994.
**Author's unpublished data.

selenium, vanadium, boron, silicon, and a few others, of which biological functions have not yet been fully recognized.

Actually, mineral deficiency states are more likely to occur than are vitamin insufficiency states. At increased risk of mineral deficiencies are people who eat low-calorie diets, the elderly, pregnant women, people using certain drugs such as diuretics, vegetarians, and those living in areas where soils are deficient in certain minerals. There is increasing evidence that those humans whose nutritional status is suboptimal in certain trace elements, such as selenium, may be at greater risk for certain forms of cancer and heart disease. Suboptimal intake can be due to soil depletion, the effects of acid rain, and the overrefining and overprocessing of foods and other factors.

Minerals occur in foods in many chemical forms. They are absorbed from the intestines as simple cations, as part of anionic groups, or in covalent or noncovalent associations with organic molecules. The chemical form of minerals in foods strongly influences their intestinal handling and biological availability. Thus, the iron in the form of hemoglobin in meats is more bioavailable than inorganic iron. This may also be true for selenium in selenomethionine and for the organic chelates of dietary chromium and zinc. Factors that affect mineral solubility or their reduction to a suitable form for cellular uptake or those that influence the transfer through the mucosa or transport into circulation govern the rate and efficiency of uptake of the minerals. For example, iron and zinc are much more bioavailable from human breast milk than from cow's milk of comparable infant formula. The intrinsic molecular associations of these minerals with low-molecular-weight binding compounds in human mammary secretions is thought to convey this enhanced absorbability (Rosenberg and Solomons, 1984). Some minerals can produce chronic toxicity when absorbed and retained in excess of the body's demands. Homeostatic mechanisms, often hormonally mediated, regulate the absorption of certain minerals and thereby protect against excessive accumulation.

Recently developed speciation analysis makes it possible to determine the forms of minerals that are present in food and in the environment and that may cause specific physiological or pharmacological effects in organisms.

INTERACTION WITH DIETARY COMPONENTS

EFFECT ON ABSORPTION

Various nutritional and nonnutritional components of the diet, other nutrients in vitamin-mineral supplements, or assorted medications can interact with minerals in the gastrointestinal tract and influence their absorption. For example, amino acids may perform as intraluminar binders for some trace minerals. Large, complex, and poorly digestible proteins, on the other hand, may bind minerals tightly and diminish their absorption. Triacylglycerols and long-chain fatty acids derived from triacylglycerols may form soaps with calcium and magnesium and decrease the bioavailability of these two nutrients.

Lactose has been implicated in the enhanced absorption of calcium from milk. Pectins, cellulose, hemicellulose, and polymers produced by the Maillard reaction during cooking, processing, or storage may bind minerals in the lumen and thus reduce their biological availability. Interaction between and among minerals or with anionic species is an important determinant of mineral absorption. Absorption of iron is hindered by fiber and phosphates and promoted by ascorbic acid, copper, and meat protein. Ascorbic acid also enhances absorption of selenium but reduces the absorption of copper. A high protein intake appears to increase the excretion of calcium, whereas vitamin D ingestion promotes the retention of calcium. Intestinal parasites, dietary fiber phytates, and excessive sweating interfere with zinc absorption. Phytates, oxalates, and tannates can interfere with the absorption of a number of minerals. Certain medications such as tetracycline can also inhibit absorption of minerals, while others such as didoquin or dilatin may actually promote uptake of certain minerals.

Apparently, chemically similar minerals share certain "channels" for absorption, and the simultaneous ingestion of two or more such minerals will result in competition for absorption. When unphysiological imbalances among competitive nutrients exist as the result of leaching from water pipes, storage in unlacquered tin cans, or improper formulation of vitamin-mineral supplements, nutritionally important consequences of mineral–mineral interaction can result. Finally, to participate in a nutritionally relevant process for the organisms as a whole, a mineral must be transported away from the intestine. The concentration of circulating binding proteins and the degree of saturation of their metallic binding sites may influence the rate and magnitude of transport of recently absorbed minerals (Rosenberg and Solomons, 1984).

Minerals require a suitable mucosal surface across which to enter the body. Resection or diversion of a large portion of small bowel obviously affects mineral absorption. Extensive mucosal damage due to mesenteric infarction or inflammatory bowel disease or major diversion by jejunoileal bypass procedures reduce the available surface area. Minerals whose absorption primarily occurs in the proximal intestine, e.g., copper or iron, are affected differently than those absorbed more distally, e.g., zinc. In addition, the integrity of the epithelium, the uptake of fluid and electrolyte, the intracellular protein synthesis, energy-dependent pumps, and hormone receptors must be intact. Intrinsic diseases of the small intestinal mucosa may impair mineral absorption. Such conditions as celiac sprue, dermatitis herpetiformis, infiltrative lymphomas, and occasionally inflammatory bowel disease produce diffuse mucosal damage. Protein energy malnutrition causes similar damage, and tropical enteropathy affects part of the population of developing countries living under adverse nutritional and hygienic conditions.

As can be seen from the above, the absorption of most metals from the gastrointestinal tract is variable (Table 3.2) and depends on many external and internal factors. Thus, the quantity of metal ingested rarely reflects that which is bioavailable. In fact, under most circumstances, only a small fraction of ingested metals is absorbed, while the great majority passes out of the gut in the feces.

The Recommended Dietary Allowances (RDA) represent the standard of nutrition set by the Food and Nutrition Board of the U.S. National Academy of Sciences (Feltman, 1990). It contains the levels of essential nutrients that are adequate to meet the nutritional needs of the normal healthy population. Individuals may differ in their precise nutritional requirements. To take into account these differences among normal persons, the RDA provides a "margin of safety"; that is, they set the allowances high enough to cover the needs of most healthy people. For additional nutrients that are necessary to keep the body in health for which RDA has not yet been established, there are estimated "safe and adequate daily intake" (SAI).

The functions of minerals in the body involve building of the tissue and regulating numerous body processes. Their role in the human body is summarized in Table 3.3.

BUILDING BODY TISSUE AND REGULATING BODY PROCESSES

Certain minerals, including calcium, phosphorus, magnesium, and fluorine are components of bone and teeth. Deficiencies during the growing years cause growth to be stunted and bone tissue to be of poor quality. A continual adequate intake of minerals is essential for the maintenance of skeletal tissue in adulthood. Potassium, sulphur, phosphorus, iron, and many other minerals are also structural components of soft tissue (Solomons, 1984; Eschleman, 1984).

Minerals are integral parts of many hormones, enzymes, and other compounds that regulate biochemical functions in the organism. For example, iodine is required to produce the hormone thyroxine, chromium is involved in the production of insulin, and hemoglobin is an iron-containing compound. Hence, the production of these substances in the organism de-

TABLE 3.2. **Mean Daily Intake and Recommended Dietary Allowances (RDAs) or Safe Adequate Intake (SAI), as Well as Percentage of Absorption of Minerals from the Gastrointestinal Tract According to Published Data.**

| Mineral | Milligram Per Adult Person | | Percentage of Absorption |
	Daily Intake	RDAs or SAI	
Macroelements			
Calcium	960–1220	800–1200*	10–50
Chloride	1700–5100	750**	high[†]
Magnesium	145–358	280–350*	20–60
Phosphorus	1670–2130	800–1200*	high[†]
Potassium	3300	2000**	high[†]
Sodium	3000–7000	500**	high[†]
Microelements			
Chromium	<0.15	0.05–0.20**	<1 or 10–25 in form of GTF[‡]
Cobalt	0.003–0.012	0.002[§]	30–50
Copper	2.4	1.5–3.0**	25–60
Fluorine	<1.4	1.5–4**	high[†]
Iodine	<1.0	0.15*	100
Iron	15	10–15*	10–40
Manganese	5.6; 8	2–3**	40
Molybdenum	>0, 15	0.075–0.250**	70–90
Nickel	0.16–0.20	0.05–0.3	<10
Selenium	0.06–0.22	0.055–0.070	~70
Vanadium	0.012–0.030	0.01–0.025	<1
Zinc	12; 18	12–15*	30–70
Microelements recently considered as essential			
Boron	1–3	1–2	high[†]
Silicon	21–46; 200	21–46	3; 40

*RDA
**SAI
[†]More than 40%.
[‡]Glucose tolerance factor.
[§]This is 0.002 mg cobalt containing vitamin B_{12}.

TABLE 3.3. The Biological Role of Some Minerals.

Mineral	Function	Deficiency	Sources
Macroelements			
Calcium	Bone and tooth formation; blood clotting, cell permeability, nerve stimulation, muscle contraction, enzyme activation	Stunted growth, rickets, osteomalacia, osteoporosis, tetany	Milk, hard cheese, salmon and small fish eaten with bones, some dark green vegetables, legumes
Magnesium	Component of bones and teeth, activation of many enzymes, nerve stimulation, muscle contraction	Seen in alcoholism or renal disease, tremors leading to conclusive seizures	Green leafy vegetables, nuts, whole grains, meat, milk, seafood
Phosphorus	Bone and tooth formation, energy metabolism—component of ATP and ADP, protein synthesis—component of DNA and RNA, fat transport, acid-base balance, enzyme formation	Stunted growth, rickets	Milk, meats, poultry, fish, eggs, cheese, nuts, legumes, whole grains
Potassium	Osmotic pressure, water balance, acid-base balance, nerve stimulation, muscle contraction, synthesis of protein, glycogen formation	Nausea, vomiting, muscular weakness, rapid heart beat, heart failure	Meats, fish, poultry, whole grains, fruits, vegetables, legumes

(continued)

TABLE 3.3. (continued).

Mineral	Function	Deficiency	Sources
Sodium	Osmotic pressure, water balance, acid-base balance, nerve stimulation, muscle contraction, cell permeability	Rare: nausea, vomiting, giddiness, exhaustion, cramps	Table salt, salted foods, MSG and other sodium additives, milk, meat, fish, poultry, eggs, bread
Microelements			
Chromium	Trivalent chromium increases glucose tolerance and plays role in lipid metabolism. Useful in prevention and treatment of diabetes. Hexavalent chromium is toxic.	Impaired growth, glucose intolerance, elevated blood cholesterol	Whole-grain cereals, condiments, meat products, cheeses and brewer's yeast
Cobalt	Cofactor of vitamin B_{12}, plays role in immunity	Rarely observed; if exists, a pernicious anemia with hematological and neuralgic manifestations; may be observed due to vitamin B_{12} deficiency	Organ meats (liver, kidney), fish, dairy products, eggs
Copper	Necessary for iron utilization and hemoglobin formation, is constituent of cytochrome oxidase, and involved in bone and elastic tissue development	Anemia, neutropenia, leucopenia, skeletal dimineralization	Liver, kidney, oysters, nuts, fruits, and dried legumes

(continued)

TABLE 3.3. (continued).

Mineral	Function	Deficiency	Sources
Iron	Hemoglobin and myoglobin, essential component of many enzymes	Anemia, decrease in oxygen transport and cellular immunity, muscle weakness	Liver, lean meats, legumes, dried fruits, green leafy vegetables, whole grain and fortified cereals
Manganese	Cofactor of large number of enzymes. In aging process has a role as an antioxidant (Mn-superoxide dismutase). Important for normal brain function, for reproduction, and for bone structure	In animals: chondrodystrophy, abnormal bone development, reproductive difficulties. In humans: shortage of evidence	Tea, whole grain, and nuts. Moderate levels: fruits and green vegetables. Organ meat and shellfish contain well absorbable manganese
Molybdenum	Cofactor of the enzymes: xanthine and aldehyde oxidase; copper antagonist	Reduces conversion of hypoxanthine and xanthine to uric acid, resulting in the development of xanthine renal calculi. Deficiency state may be potentiated by high copper intake.	Grain, legumes
Zinc	Consistent of many enzyme systems, carbon dioxide transport, and vitamin A utilization	Delayed wound healing, impaired taste sensitivity, retarded growth and sexual development, dwarfism	Oysters, fish, meat, liver, milk, whole grains, nuts, legumes

(continued)

44

TABLE 3.3. (continued).

Mineral	Function	Deficiency	Sources
Fluorine	Resistance to dental decay	Tooth decay in young children	Drinking water rich in fluoride, seafood, teas
Iodine	Synthesis of thyroid hormones that regulate basal metabolic rate	Goiter; cretinism, if deficiency is severe	Iodized salt, seafood, food grown near the sea
Selenium	Protects against number of cancers	Cataract, muscular dystrophy, growth depression, liver cirrhosis, infertility, cancer, aging due to deficiency of selenoglutathione peroxidase, insufficiency of cellular immunity	Broccoli, mushrooms, radishes, cabbage, celery, onions, fish, organ meats
Boron (recently considered as essential)	Prevents osteoporosis in postmenopausal women, beneficial in treatment of arthritis, builds muscle. Its essentiality has still to be proven.	Probably impairs growth and development	Foods of plant origin and vegetables

Sources: Eschleman, 1984; Hendler, 1990.

pends upon adequate intake of the involved minerals. Calcium, for example, is a catalyst in blood clotting. Some minerals act as catalyst in the absorption of nutrients from the gastrointestinal tract in the metabolism of proteins, fat, and carbohydrates and in the utilization of nutrients by the cell.

Minerals dissolved in the body fluids are responsible for nerve impulses and the contraction of muscles, as well as for water and acid-base balance. They play important roles in maintaining the respiration, heart rate, and the blood pressure within normal limits. Deficiency of minerals in the diet may lead to severe, chronic, clinical signs of diseases, frequently reversible after their supplementation in the diet or following the total parenteral nutrition. Their influence on biochemical reactions in living systems also makes it possible to use them intentionally in many food processes.

ROLE IN FOOD PROCESSING

EFFECT ON OXIDATION

Because minerals are an integral part of many enzymes, they thus play an important role in food processing, e.g., in alcoholic and lactic fermentation, meat aging, and in dairy food production. Many compounds used as food additives or for rheological modification of some foods also contain metallic cations in their structure. A number of these compounds function as antimicrobial, sequestrant, antioxidant, flavor enhancer, and buffering agents and sometimes as dietary supplements (Table 3.4).

Some heavy metal ions actively catalyse lipid oxidation. Their presence, even in trace amounts, has long been recognized as potentially detrimental to the shelf life of fats, oils, and fatty foods. They can activate molecular oxygen by producing superoxide, which then, through dismutation and other steps of biochemical changes, turns into hydroxyl radicals. Three cations are involved in the activity of superoxide dismutases (SOD). This enzyme has been patented as an antioxidant agent for foods. Three types of metalloenzymes of SOD exist in living organisms, namely Cu Zn-SOD, Fe-SOD, and Mn-SOD. All three type of SODs catalyze dismutation of superoxide anions to produce hydrogen peroxide in vivo. There is evidence of increased lipid oxidation in apple fruit during senescence. SOD activity may also be involved in reactions to damage induced by oxygen, radicals, and ionizing radiation and could help to protect cells from damage by peroxidation products (Du and Bramlage, 1994).

Besides SOD, catalase, ceruloplasmin, albumin and appotransferrin, and chelating agents [e.g., ethylenediaminetetraacetic acid (EDTA)], bathocupreine, cysteine, and purine are capable of inhibiting the oxidation of ascorbic acid induced by trace metals. Copper-induced lipid oxidation

TABLE 3.4. List of Selected Mineral Compounds Used as Food Additives According to FAO/WHO Expert Committee on Food Additives (1994).

Chemical Name of Compound and (INS)*	Synonyms or Other Chemical Name	Functional Class and Comments	ADI, TADI, PMTDI,** (mg/kg body weight)
Calcium alginate (404)	Calcium alginate	Thickening agent, stabilizer	ADI "not specified"
Calcium ascorbate (302)	Calcium ascorbate dihydrate	Antioxidant	ADI "not specified"
Calcium benzoate (213)	Monocalcium benzoate	Antimicrobial, preservative	ADI 0–5.0
Calcium chloride (509)	Calcium chloride	Firming agent	ADI "not specified"
Calcium citrate (333)	Tricalcium citrate, tricalcium salt of beta hydroxytricarballylic acid	Acidity, regulator, firming agent, sequestrant	ADI "not specified"
Calcium dihydrogen phosphate (341i)	Calcium dihydrogen tetra oxophosphate. Monobasic calcium phosphate. Monocalcium phosphate	Buffer, firming, raising, leavening and texturing agent, and in fermentation process	MTDI 70.0
Calcium disodium ethylenediaminetetraacetate (385)	Calcium disodium EDTA	Antioxidant, preservative, sequestrant (No excess of disodium EDTA should remain in food)	ADI 0–2.5
Calcium glutamate (623)	Monocalcium dl-L-glutamate	Flavor enhancer, salt substitute	ADI "not specified"
Calcium hydroxide (526)	Slaked lime	Neutralizing agent, buffer, firming agent	ADI "not specified"
Calcium hydrogen carbonate (170ii)	Calcium hydrogen carbonate	Surface colorant, anticaking agent, stabilizer	ADI "not specified"
Calcium lactate (327)	Calcium dilactate hydrate	Buffer, dough conditioner	
Calcium sorbate (203)	Calcium sorbate	Antimicrobial, fungistatic, preservative agent	ADI 0–25.0 (as sum of calcium, potassium and sodium salt)
Magnesium chloride (511)	Magnesium chloride hexahydrate	Firming, color retention agent	ADI "not specified"

(continued)

47

TABLE 3.4. (continued).

Chemical Name of Compound and (INS)*	Synonyms or Other Chemical Name	Functional Class and Comments	ADI, TADI, PMTDI,** (mg/kg body weight)
Magnesium carbonate (504i)	Magnesium carbonate	Anticaking and antibleaching agent	ADI 0–50.0
Magnesium gluconate (580)	Magnesium gluconate dihydrate	Buffering, firming agent, yeast food	ADI "not specified"
Magnesium glutamate dl-L- (625)	Magnesium glutamate	Flavor enhancer, salt substitute	ADI "not specified" (group ADI for α glutamic acid and its mono-sodium, potassium, calcium, magnesium and ammonium salts)
Magnesium hydrogen phosphate (343ii)	Magnesium hydrogen or-thophosphate trihydrate. Dimagne-sium phosphate	Dietary supplement	MTDI 70 (expressed as phosphorus from all sources)
Magnesium hydroxide (528)	Magnesium hydroxide	Alkali, color adjunct	ADI "not limited"
Magnesium hydroxide car-bonate (504ii)	Magnesium carbonate hydroxide hydrated	Alkali, anticaking, color retention, carrier, drying agent	ADI "not specified"
Magnesium lactate D,L- also magnesium lactate L (329)	Magnesium D,L-lactate	Buffering agent, dough condi-tioner, dietary supplement	ADI "not limited"
Magnesium oxide (530)	Magnesium oxide	Anticaking, neutralizing agent	ADI "not limited"
Magnesium sulfate (518)	Magnesium sulfate	Firming agent	ADI "not specified"
Potassium acetate (261)	Potassium acetate	Antimicrobial, preservative, buffer	ADI "not specified" (also includes the free acid)

(continued)

TABLE 3.4. (continued).

Chemical Name of Compound and (INS)*	Synonyms or Other Chemical Name	Functional Class and Comments	ADI, TADI, PMTDI,** (mg/kg body weight)
Potassium alginate (402)	Potassium alginate	Thickening agent, stabilizer	ADI "not specified" (group ADI for alginic acid and its ammonium, calcium, and sodium salts)
Potassium aluminosilicate (555)	Potassium aluminosilicate	Anticaking agent	No ADI allocated
Potassium ascorbate (303)	Potassium ascorbate	Antioxidant	ADI "not specified" (group ADI for ascorbic acid and its sodium, potassium, and calcium salts)
Potassium benzoate (212)	Potassium benzoate	Antimicrobial, preservative	ADI 0–5.0 (expressed as benzoic acid)
Potassium bromate (924a)	Potassium bromate	Oxidizing agent	ADI withdrawn
Potassium carbonate (501i)	Potassium carbonate	Alkali, flavor	ADI "not specified"
Potassium chloride (508)	Potassium chloride, sylvine, sylvite	Seasoning and gelling agent, salt substitute	ADI "not specified" (group ADI for hydrochloric acid and its magnesium, potassium, and ammonium salts)
Potassium or sodium copper chlorophyllin (141ii)	Potassium or sodium chlorophyllin	Color of porphyrin	ADSI 0–15
Potassium dihydrogen phosphate (340i)	Monopotassium dihydrogen ortho-phosphate, monobasic potassium-phosphate	Buffer, sequestrant, neturalizing agent	MTDI 70.0

(continued)

TABLE 3.4. (continued).

Chemical Name of Compound and (INS)*	Synonyms or Other Chemical Name	Functional Class and Comments	ADI, TADI, PMTDI,** (mg/kg body weight)
Potassium hydrogen carbonate (501ii)	Potassium bicarbonate	Alkali, leavening agent, buffer	ADI "not specified"
Potassium hydrogen sulfite (228)	Potassium hydrogen sulfite	Preservative, antioxidant	ADI 0–0.7 (group ADI for sulfur dioxide and sulfites, expressed as sulfur dioxide, covering sodium and potassium metabisulfite, potassium, and sodium hydrogen sulfite and sodium thiosulfate)
Potassium glutamate (622)	L-Monopotassium L-glutamate	Flavor enhancer, salt substitute	ADI "not specified"
Sodium alginate (401)	Sodium alginate	Thickening agent, stabilizer	ADI "not specified"
Sodium aluminium phosphate acidic (541i)	Salp. sodium trialuminium tetradecahydrogen, octaphosphate tetrahydrate (A); trisodium di-aluminium pentadecahydrogen octaphosphate (B)	Raising agent	ADI 0–0.6
Sodium aluminium phosphate basic (541ii)	Kasal; autogeneous mixture of an alkaline sodium aluminium phosphate	Emulsifier	ADI 0–0.6
Sodium ascorbate (301)	Sodium L-ascorbate	Antioxidant	ADI "not specified"
Sodium benzoate (211)	Sodium salt of benzenecarboxylic acid	Antimicrobial, preservative	ADI 0–5.0
Sodium dihydrogen phosphate (339)	Monosodium dihydrogen monophosphate (orthophosphate)	Buffer, neutralizing agent, sequestrant in cheese, milk, fish, and meat products	MTDI 70.0

(continued)

TABLE 3.4. (continued).

Chemical Name of Compound and (INS)*	Synonyms or Other Chemical Name	Functional Class and Comments	ADI, TADI, PMTDI,** (mg/kg body weight)
Disodium ethylenediamine tetraacetate (386)	Disodium EDTA; disodium edeteate	Antioxidant, sequestrant, preservative, synergist	ADI 0–2.5 (as calcium disodium EDTA)
Sodium glutamate (621)	Monosodium L-glutamate, (MSG); glutamic acid monosodium salt monohydrate	Flavor enhancer	ADI "not specified"
Sodium iron III—ethylenediamine tetraacetate trihydrate	Ferric sodium edeteate, sodium iron EDTA, sodium feredetate	Nutrient supplement (provisionally considered to be safe in food fortification programs)	ADI acceptable
Sodium or potassium metabisulfite (223, 224)	Disodium or potassium pentaoxodisulfate	Antimicrobial, preservative, bleaching agent, antibrowning agent	ADI 0–0.7 (group ADI for sulphur dioxide and sulfites expressed as SO_2, covering sodium and potassium salt)
Sodium nitrite (250)	Sodium nitrite	Antimicrobial, color fixative	ADI 0–0.2
Sodium nitrate (251)	Sodium nitrate, cubic or soda nitre, chile salpetre	Antimicrobial, color fixative	ADI 0–5.0

(continued)

TABLE 3.4. (continued).

Chemical Name of Compound and (INS)*	Synonyms or Other Chemical Name	Functional Class and Comments	ADI, TADI, PMTDI,** (mg/kg body weight)
Sodium phosphate (339iii)	Trisodium phosphate; trisodium monophosphate, orthophosphate; sodium phosphate	Sequestrant, emulsion stabilizer, buffer	MTDI 70.0
Sodium or potassium sorbate (201, 202)	Sodium or potassium sorbate	Antimicrobial, fungistatic agent	ADI 0–25.0

*INS—International numbering system has been prepared by the Codex Committee for Food Additives for the purpose of providing an agreed international numerical system for identifying of food additives in ingredient list as an alternative to the declaration of the specific name (*Codex Alimentarius*, vol. 1, Second Edition, Section 5.1, 1992).

**ADI—Acceptable Daily Intake is an estimate of the amount of a substance in food or drinking water, expressed on a body weight basis, for a standard human of 60-kg weight, that can be ingested daily over a lifetime without appreciable risk for health.

ADI "not specified" or ADI "not limited"—terms applicable to a food substance of very low toxicity which, on the basis of the available data—chemical, biochemical, toxicological, and other—as well as that of the total dietary intake of the substance arising from its use at the levels to achieve the desired effect and from its acceptable background in food, does not, in the opinion of the Joint FAO/WHO Expert Committee on Food Additives (JECFA), represent a hazard to health. For that reason and for reasons stated in individual evaluations, the establishment of ADI in numerical form is not deemed necessary. An additive meeting this criterion must be used within the bound of good manufacturing practice; i.e., it should be technologically efficacious and should be used at the lowest level necessary to achieve this effect; it should not conceal inferior food quality or adulteration, and it should not create a nutritional imbalance.

TADI—Temporary ADI is a term established by the JECFA for substances for which toxicological data are sufficient to conclude that use of the substance is safe over the relatively short period of time required to evaluate further safety data, but are insufficient to conclude that use of the substance is safe over a lifetime. A higher than normal safety factor is used when establishing a TADI, and an expiration date is established by which time appropriate data to resolve the safety issue should be submitted to JECFA.

MTDI—Maximum Tolerable Daily Intake, or Provisional Maximum Tolerable Daily Intake (PMTDI) is a term used for description of the endpoint of contaminants with no cumulative properties. Its value represents permissible human exposure as a result of the natural occurrence of the substance in food or drinking water. In the case of trace elements that are both essential nutrients and unavoidable constituents of food, a range is expressed, the lower value representing the level of essentiality and the upper value the PMTDI.

of ascorbic acid induced by trace metals. Copper-induced lipid oxidation in ascorbic acid pretreated cooked ground fish may be inhibited in the presence of natural polyphenolic compounds, the flavonoids, which are effective antioxidants and prevent the production of radicals (Ramanathan and Das, 1993). In the presence of ADP-chelated iron and traces of copper, oxygen radicals are generated in the sarcoplasmic reticulum of muscle food. Muscle contains notable amounts of iron, a known prooxidant, and trace amounts of copper, which also catalyze peroxidative reaction (Hultin, 1994; Wu and Brewer, 1994). Iron occurs associated with heme compounds and as nonheme iron complexed to proteins of low-molecular-weight. Reactive nonheme iron can be obtained by release of iron from heme pigments or from the iron storage protein, ferritin. Iron is part of the active site of lipoxygenase, which may participate in lipid oxidation. Reducing components of the tissue-like superoxide anion, ascorbate and thiols can convert the inactive ferric iron to the active ferrous iron. There are also enzymic systems that use reducing equivalents from NADPH to reduce ferric iron. A number of cellular components are capable of reducing ferric to ferrous iron, but under most conditions, the two major reductants are superoxide and ascorbate (Hultin, 1994). In some cases, reduction of ferric iron can also be accomplished enzymically utilizing electrons from NADH and, to a lesser extent, NADPH by enzymic system associated with both the sarcoplasmic reticulum and mitochondria. Ferrous iron can activate molecular oxygen by producing superoxide. Superoxide may then undergo dismutation spontaneously or by the action of SOD and produce hydrogen peroxide, which can interact with another atom of ferrous iron to produce the hydroxyl radical. The hydroxyl radical can initiate lipid oxidation. It is generally accepted that ferrous iron is the reactive form of iron in oxidation reaction. Since it is likely that most iron ordinarily exists in the cell as ferric iron, the ability to reduce ferric to ferrous iron is thus critical.

Development of rancidity and warmed-over flavor—a specific defect that occurs in cooked, reheated meat products following short-term refrigerated storage—has been directly linked to autoxidation of highly unsaturated, membrane-bound phospholipids and to the catalytic properties of nonheme iron (Oellingrath and Slinde, 1988; Pearson et al., 1977; Hultin, 1994).

Dietary iron may influence muscle iron stores and thus theoretically may also affect the lipid oxidation in muscle food, e.g., pork. There appears to be a threshold for dietary iron level (between 130–210 ppm), above which muscle and liver nonheme iron and total iron, and muscle thiobarbituric acid reactive substances began to increase because of porcine muscle lipid oxidation (Miller, Smith, et al., 1994; Miller, Gomez-Basauri, et al., 1994).

The secondary oxidation products, mainly aldehydes, are the major contributors to warmed-over flavor and meat flavor deterioration, because of their high reactivity, and low flavor thresholds. Ketones and alcohols have a high flavor threshold and are lesser causes of off-flavors.

Exogenous antioxidants can preserve the quality of meat products. Radical scavengers appear to be the most effective inhibitors of meat flavor deterioration. However, different substrates and systems respond in different ways. Active ferrous iron may be eliminated physically by chelation with EDTA or phosphates or chemically by oxidation to its inactive ferric form.

Ferroxidases are enzymes that oxidize ferrous to ferric iron in the presence of oxygen according to the formula:

$$4Fe^{2+} + O_2 + 4H^+ \rightarrow 4Fe^{3+} + 2H_2O$$

Ceruloplasmin, a copper protein of blood serum, is a ferroxidase. Oxidation of ferrous to ferric iron tends to be favored in extracellular fluids, while chelation is more likely intracellularly (Hultin, 1994).

Sodium chloride has long been used for food preservation. Salt alters both the aroma and the taste of food. Addition of sodium chloride to blended cod muscle accelerates the development of rancidity (Castell et al., 1965; Castell and Spears, 1968). This salt-induced rancidity is inhibited by chelating agents such as EDTA, sodium oxalate, and sodium citrate and by the nordihydroguaiaretic acid and propyl gallate. Although sodium chloride and other metal salts act as prooxidants, they have a strong inhibiting effect on Cu^{2+}-induced rancidity in the fish muscle. The most effective concentration of NaCl for this antioxidant effect is between 1% and 8% (Castell et al., 1965; Castell and Spears, 1968). Castell and Spears (1968) also showed that the other heavy metal ions were effective in producing rancidity when added to various fish muscles. The relative effectiveness was of the following decreasing order: $Fe^{2+} > V^{2+} > Cu^2 > {}^+Fe^{3+} > Cd^{2+} > Co^{2+} > Zn^{2+}$, while $Ni,^{2+}Ce^{2+}$, Cr^{3+} and Mn^{2+} had no effect in the used concentrations. Of those tested, Fe^{2+}, V^{2+}, and Cu^{2+} were by far the most active catalysts. There were, however, important exceptions. The comparative effectiveness of the metallic ions was not the same for muscle taken from all the species that were tested.

EDTA is reported to be effective as the metal ion sequestrant and is approved for use in the food industry as a stabilizer and antioxidant. It also acts as inhibitor of *Staphylococcus aureus* by forming stable chelates in the media with multivalent cations, which are essential for cell growth. The effect is largely bacteriostatic and easily reversed by releasing the complexed cations with other cations for which EDTA has higher affinity (Kraniak and Shelef, 1988).

The addition of phosphate — pyro-, tripoly-, and hexametaphosphate also protects cooked meat from autoxidation. Orthophosphate gives no protection. The mechanism by which phosphates prevent autoxidation appears to be related to their ability to sequester metal ions, particularly ferrous iron, which are the major prooxidants (Pearson et al., 1977). The addition of NaCl increases retention of moisture in meat and meat products.

EFFECT ON RHEOLOGICAL PROPERTIES

The interaction between metal ions and polysaccharides often affects the rheological and functional properties in food systems (Ha et al., 1989).

In aqueous media, neutral polysaccharides have little affinity for alkali metal and alkali earth metal ions. On the other hand, anionic polysaccharides have a strong affinity for metal counterions. This association is related to the linear charge density of the polyanions. The linear charge density is expressed by the distance between the perpendicular projections of adjacent charged groups on the main axis of the molecule. The higher the linear charge density, the stronger the interaction of counterions with anionic groups of the molecule. Such anionic hydrocolloids (0.1% solutions) as alginate, karaya, arabic, and ghati have higher calcium-binding affinity (Ha et al., 1989). An important functional property of alginates is their capacity to form gels with calcium ions. This makes alginates extensively suited to prepare products such as fruit and meat analogs. They are also widely used in biotechnology as an immobilization agent of cells and enzymes. The method involves diffusion of calcium ions through alginate and a cross-linking reaction with alginate carboxylic group to form the gel (Ha et al., 1989; WHO FAS 30, 1993; WHO FAS 32, 1993). Carrageenans are reported to stabilize casein and several plant proteins against precipitation with calcium and are used to prepare texturized milk products (Samant et al., 1993).

OTHER EFFECTS

Sodium reduction in diet is recommended as a means of preventing hypertension and subsequent cardiovascular disease, stroke, and renal failure. Reducing or substituting NaCl requires an understanding of the effects caused by the new factors introduced. Several ways are proposed to reduce the sodium content in processed meat without adverse effect on the quality (flavor, gelation, etc.) and the shelf life of the products. This includes a slight sodium chloride reduction, replacing some of the NaCl with other chloride salt (KCl, $MgCl_2$), or nonchloride salt and/or altering processing methods (Barbut and Mittal, 1985).

Calcium ion is a known activator of many biochemical processes. The calcium activated neutral protease (CANP) plays an important role in post-mortem tenderizing of meat. The function of the metal ion in such an enzyme is believed to be either neutralization of the charges on the surface by preventing electrostatic repulsion of subunits or effecting of a conformational change required for association of the subunits. Thus, the metal ions must be present in a specific state to perform this function.

The metallic cations in solution exist as aqua-complex ions in equilibrium with their respective hydroxy-complex:

$$M(H_2O)^{m+} \longleftrightarrow MOH^{(m-1)+} + H^+$$

aqua-complex ion	hydroxy-complex
	(weak base)

The acid ionization constant (pK_a) of the aqua-complex ion determines whether or not the ion would form complexes with a protein. This depends greatly on the pH of the medium. Since the ionization constant of low charge is 12.6, they would form a stable complex only with negatively charged protein in alkaline media. They cannot bind to cationic proteins because they do not share electrons to form a covalent bond. This consideration explains why the activity of Ca^{2+}-activated protease is optimum in the alkaline pH range. Thus, a decrease in its activity at acidic pH values may partly be due to a change in the electronic state of Ca^{2+} (Asghar and Bhatti, 1987; Barbut and Mittal, 1985).

Generally, sodium and potassium react only to a limited extent with proteins, whereas calcium and magnesium are somewhat more reactive. Transition metals, e.g., ions of Cu, Fe, Hg, and Ag, react readily with proteins, many forming stable complexes with thiol groups. Calcium cations and ferrous and cupric, as well as magnesium cations may be an integral part of certain protein molecules or molecular associations. Their removal by dialysis or by sequestrants appreciably lowers the stability of the protein structure toward heat and proteases.

THE EFFECT OF STORAGE AND PROCESSING ON THE MINERAL COMPONENTS IN FOODS

The effect of normal storage on mineral components is rather low and may be connected mainly with changes of humidity or contamination; however, high changes of mineral components may occur during canning, cooking, drying, freezing, peeling, and all the other steps involved in preserving, as well as in food processing for direct consumption.

The highest losses of minerals are encountered in the milling and polishing process of cereals and groats. All milled cereals undergo a significant reduction of nutrients. The extent of the loss is governed by the efficiency with which the endosperm of the seed is separated from the outer seed coat-bran and the germ. The loss of certain minerals and vitamins is deemed so relevant to health that, in many countries, a supplementation procedure was introduced for enriching food products with the lost nutrients, i.e., iron to the bread. In some countries, regulations were issued for standards of identity for enriched bread. If bread is labeled "enriched," it must meet these standards. In white flours, the losses of magnesium and manganese may reach up to 90%. These minerals remain mainly in the bran—the outer part of the cereals. For this reason, it seems reasonable to recommend consumption of bread baked from whole meals instead of from white meals. Recommended, sometimes steady, consumption of bran alone, for dietary purposes, should be done with great care because it may also contain many different contaminants, like toxic metals and organic pesticides.

During preparation for cooking or for canning, vegetables should be thoroughly washed before cutting to remove dirt and traces of insecticide spray. Root vegetables should be scrubbed. The dark outer leaves of greens are rich in iron, calcium, and vitamins, so they should be trimmed sparingly. Peeling vegetables and fruit should be avoided whenever possible, because minerals and vitamins are frequently concentrated just beneath the skin. Potatoes should be baked or cooked in their skins, even for hash browns or potato salad. True et al. (1979) showed that cooking potatoes by boiling whole or peeled tubers, as well as microwave cooked and oven baked, may have a negligible effect on the losses of Al, B, Ca, Na, K, Mg, P, Fe, Zn, Cu, Mn, Mo, J, and Se. Microwaved potatoes retain nutrients well, and contrary to popular belief, peeling potatoes does not strip away their vitamin C and other minerals. Whenever practical, any remaining cooking liquid should be served with the vegetable or used in a sauce or gravy soup. To retain minerals in canned vegetables, one should pour the liquid from the can into a saucepan, and heat at low temperature to reduce liquid, add vegetables to remaining liquid and heat before serving. Low temperatures reduce shrinkage and loss of many other nutrients. Cooking and blanching leads to the most important nutrient losses. To the liquid of the cooked vegetables is leached 30–65% of potassium, 15–70% of magnesium and copper, and from 20% to over 40% of zinc. Thus, it is reasonable to use this liquid for soup preparation (Rutkowska, 1975; Trzebska-Jeske et al., 1973).

The losses depend both on the kind of vegetables cooked and the course of the applied process. Steam blanching generally results in smaller losses of nutrients, since leaching is minimized in this process. Frozen meat and vegetables thawed at ambient temperature lose many nutrients, including minerals, in the thaw drip. To avoid the losses of these nutrients, the drip should be added to the pot where the meal is prepared for consumption.

Frozen fruits should be eaten without delay, fresh, just after thawing, together with the secreted juice. Foods blanched, cooked, or reheated in a microwave oven generally retain about the same, or even higher, amounts of nutrients as those cooked by conventional methods.

THE CHEMICAL NATURE OF TOXICITY OF SOME MINERAL FOOD COMPONENTS

A diet consisting of a variety of foods provides the best protection against potentially harmful chemicals in food. This is because the body tolerates very small quantities of many toxic substances but has only limited ability to cope with large quantities of any single one. Almost any chemical can have a harmful effect if taken in large quantity. This is especially true for trace minerals and, to some degree, also for macroelements, as well as for vitamins. For this reason, it is important to understand the difference between toxicity and hazard. Many foods contain toxic chemicals, but these chemicals do not present a hazard if consumed in allowable amounts.

Toxic compounds of such metals as arsenic, mercury, cadmium, and lead contaminate the environment and may enter the food supply. A number of minerals can produce chronic toxicity when absorbed and retained in excess of the body's demands. The proportion of elements accumulated by the organism is different to the proportion of the environment, and this results in concentrations within the organisms. Some of the elements are necessary to the organism for metabolic processes, though others that are accumulated in high proportion, sometimes specifically in some organs, and, according to the present knowledge, do not have any metabolic significance for the organism like arsenic, cadmium, mercury, and lead, are recognized as toxic. Their toxicity is a function of the chemical form and the dose that enters into the body. It is also a function of accumulation in the body tissues. For this reason, it is very important to have the information about the chemical form of the discussed metal. Currently, this may be done by applying "speciation analysis," which makes it possible to differentiate the chemical form of the examined element and to assess the safety level of the metal residue in foods or in drinking water. For example, pentavalent and trivalent arsenicals each react with biological ligands in different ways. The trivalent form reacts with the thiol protein groups, resulting in enzyme inactivation, structural damage, and a number of functional alterations. The pentavalent arsenicals, however, do not react with $-SH$ groups. Arsenate can competitively inhibit phosphate insertion into the nucleotide chains of DNA of cultured human lymphocytes, causing false formation of DNA because of instability of the arsenate esters. Dark repair mechanisms are also inhibited, leading to persistence of these errors in the DNA molecules. Binding difference for the trivalent and pentavalent forms probably leads to the differences in accumulation of this element. Trivalent inorganic As is accumulated in a higher level than the pentavalent form. The organic arsenic compounds are considered less toxic than inorganic arsenic of which trivalent arsenicals are the most toxic forms. Fish are the richest source of arsenic in foods and contain from 0.1% to more than 40% of inorganic form, compared to total arsenic content (Vaessen and van Ooik, 1989).

On the other hand, organic mercury compounds, especially methylmercury are recognized as more dangerous for man than inorganic ones. Because the methylation process naturally occurs in biosphere, most fish contain from about 50% to over 90% mercury as methylmercury. Marlin is the only pelagic fish known to have more than 80% of the total muscle mercury present as inorganic mercury (Cappon and Smith, 1982).

Cadmium shares chemical properties with zinc and mercury, but in contrast to mercury, it is incapable of environmental methylation due to the instability of the monoalkyl derivate. Similarities and differences also exist in the metabolism of Zn, Cd, and Hg. Metallothionein and other Cd-binding proteins hold and/or transport Cd, Zn, and Hg within the body. Cadmium, like mercury and zinc, accumulates in the liver and, presumably, the kidney by strong binding cysteine residues of metallothionein or the components of other binding proteins. The high affinity of Cd for $-SH$ groups and the ability of imparting moderate covalency in bounds result in increased lipid solubility, bioaccumulation, and toxicity. In humans, after normal levels of exposure, about 50% of the body burden is found in the kidneys, about 15% in the liver, and about 20% in the muscles. As in animals, the proportion of cadmium in the kidney decreases as the liver concentration increases. The lowest concentrations of cadmium are found in brain, bone, and fat. Accumulation in the kidney continues to 50–60 years of age in humans and falls thereafter, possibly due to age-related changes in the kidney integrity function. In contrast, the cadmium level in the muscle continues to increase over the course of life. The average cadmium concentration in the renal cortex of nonoccupationally exposed persons aged 50 varies between 11 and 100 mg/kg in different regions. A daily intake of 62 μg would be required to reach a concentration of 50 mg/kg in the renal cortex at the age of 50, assuming an absorption ratio of 5%. About 10% of the absorbed daily dose is rapidly excreted (WHO.FAS, 1989).

The Joint FAO/WHO Expert Committee on Food Additives (JECFA) and other WHO committees have recognized that infants and children are the groups of higher risk to lead exposure from food and drinking water. Lead as an antropogenic contaminant finds its way into the air, water, and surface soil. Lead containing manufactured products also contribute to the lead body burden. The domestic environment, in which infants and children spend the greater part of their time, is of particular importance as the source of lead intake. In addition to exposure from general environmental sources, some infants and young children, as a result of normal, typical behavior, can receive high doses of lead through mouthing or swallowing of nonfood items. Pica, the habitual ingestion of nonfood substances, which occurs among many young children, has frequently been implicated in the etiology of lead toxicity. In the United States, on average, two-year-old children may receive about 44% of their daily lead intake from dust, 40% from food, 15% from water and beverages, and 1% from inhaled air (WHO.FAS, 1986).

The biochemical basis of lead toxicity is its ability to bind to biologically important molecules, thereby interfering with their function by a number of mechanisms. Lead may compete with essential cations for binding sites, inhibiting enzyme activity, or alter the transport of essential cations such as calcium. At the subcellular level, the mitochondrion appears to be the main target organelle for toxic effects of lead in many tissues. Lead has been shown to selectively accumulate in the mitochondria. There is evidence that it causes structural injury to these organelles and impairs basic cellular energetics and other mitochondrial functions. It is a cumulative poison, producing a continuum of effects, primarily on the hematopoietic system, the nervous system, and the kidneys. At very low blood levels,

TABLE 3.5. The Provisional Tolerable Weekly Intake (PTWI*) or Toxic Elements Estimated by the Joint FAO/WHO Expert Committee on Food Additives for Man (1986, 1989).

Element	PTWI (μg/kg body weight)	Comments
Arsenic	15.0	For inorganic arsenic
Cadmium	7.0	—
Lead	25.0	When blood lead levels in children exceed 25 μg/100 cm^3 (in whole blood), investigations should be carried out to determine the major sources of exposure, and all possible steps should be taken to ensure that lead levels in food are as low as possible.
Mercury	3.3 as methylmercury and 5.0 as total mercury	With the exception of pregnant and nursing women who are at greater risk to adverse effects from methylmercury

*PTWI—Provisional Tolerable Weekly Intake. This term refers to the contaminants such as heavy metals with cumulative properties. Its value represents permissible human weekly exposure to those contaminants unavoidably associated with the consumption of otherwise wholesome and nutritious foods.

TABLE 3.6. Metal Toxicity in Man.

Metal	Toxic Effects	Daily Intake (mg per adult person)	Source of Exposure	Absorption (percent)
Arsenic	Inorganic compounds cause abnormal skin hyperpigmentation, hyperkeratosis, skin and lung cancer. Organoarsenic compounds present in fish are less or nontoxic.	0.0–0.29	Contaminated water, food containing a residue of arsenic pesticides, and veterinary drug. Fish and shellfish are the richest source of organic compounds, arsenobetaine and arsenocholine.	Organoarsenic compounds >90%, and inorganic trivalent compounds high
Cadmium	Accumulate mainly in liver and renal cortex. Nephrotoxicity, decalcification, osteoporosis, osteomalacia, and Itai Itai disease. In early gestation, embryotoxic. Impairs immune system, calcium and iron absorption. Hypertension and cardiovascular disease. Kidney is the critical organ.	<0.05	Oysters, cephalopods, crops growing on land fertilized with contaminated phosphate and sewage sludge; cadmium leaching from enamel and pottery glazes; contaminated water	3–10%. Well absorbed is cadmium bound to metallothionein
Lead	At blood levels greater than 40 µg/100 ml, exerts a significant effect on hemopoietic system resulting in anemia; affects central nervous system	<0.1	Food contaminated from leaching of glazes of ceramic foodware, as well as from motor vehicle exhausts, atmospheric deposits; canned food, water supply from plumbing system	5–10% in adult person, and 40–50% in children
Mercury	Methylmercury compounds easy pass the blood/brain and placetal barriers. Cause severe neurological damage, greater in young children, and renal tubular dysfunction	<0.02	Fish and shellfish, meat from animal fed with mercury dressed grains	>90% as methylmercury compounds, and 15% as inorganic mercuric compounds

Sources: Nabrzyski and Gajewska, 1984; Nabrzyski et al., 1985; WHO Environmental Health Criteria, 1981; WHO Food Addit. Ser., 1986, 1989.

61

lead may impair normal metabolic pathways in children. At least three enzymes of a heme biosynthetic pathway are affected. A level of about 10 $\mu g/cm^3$ of lead in blood interferes with δ-aminolevulinic acid dehydratase (WHO.FAS, 1986). Alteration in the activity of the enzymes of the heme synthetic pathway leads to accumulation of the intermediates of the pathway. There is some evidence that accumulation of one of the intermediates, δ-aminolevulinic acid, exerts toxic effects on neural tissues through interference with the activity of the neurotransmitter γ-amino-butyric acid. The reduction in heme production per se has also been reported to adversely affect nervous tissue by reducing the activity of tryptophan pyrollase, a heme requiring enzyme. This results in an increased metabolism of tryptophan via a second pathway, which produces high blood and brain levels of the neurotransmitter serotonin. Lead interferes with vitamin D metabolism, since it inhibits hydroxylation of 25-hydroxy-vitamin D to produce the active form of vitamin D. The effect has been reported in children at blood levels as low as 10–15 $\mu g/100$ cm^3 (WHO.FAS, 1986). Measurements of the inhibitory effects of lead on heme synthesis is widely used in screening tests to determine whether medical treatment for lead toxicity is needed for children in high-risk populations who have not yet developed overt symptoms of lead poisoning. Children absorb lead from the diet with greater efficiency than do adults (Table 3.6).

Data concerning the toxicity of the four discussed toxic minerals are presented in Tables 3.5 and 3.6. The uptake of elements is not entirely independent of one another. Elements of similar chemical properties tend to be taken up together. Sometimes, one element has an inhibiting effect on another or there can be a synergistic effect, e.g., enhancement of absorption of calcium in the presence of adequate amounts of phosphorus, or cadmium and lead hindering calcium and iron absorption, or zinc and copper antagonism and their influence on the ratio of Zn/Cu on copper deficiency.

REFERENCES

Asghar, A. and A. R. Bhatti. "Endogenous Proteolytic Enzymes in Skeletal Muscle: Their Significance in Muscle Physiology and during Postmortem Aging Events in Carcasses," *Advances in Food Res.*, 31:343.

Barbut, S. and G. S. Mittal. 1985. "Rheological and Gelation Properties of Meat Batters Prepared with Three Chloride Salts," *J. Food Sci.*, 53:1296.

Cappon, C. J. and J. C. Smith. 1982. "Chemical Form and Distribution of Mercury and Selenium in Edible Seafood," *J. Anal. Toxicol.*, 6:10–21.

Castell, C. H., J. MacLean and B. Moore. 1965. "Rancidity in Lean Fish Muscle. IV Effect of Sodium Chloride and Other Salts," *J. Fish. Res. Board of Canada*, 22:929.

Castell, C. H. and D. M. Spears. 1968. "Heavy Metal Ions and the Development of Rancidity in Blended Fish Muscle," *J. Fish. Res. Board of Canada*, 25:639.

Causeret, J. 1962. "Fish as a Source of Mineral Nutrition," in *Fish as Food, Vol. 2*. G. Borgstrom, ed. New York, London: Academic Press, pp. 205–234.

Du, Z. and W. J. Bramlage. 1994. "Superoxide Dismutase Activities in Senescin Apple Fruit (*Malus domestica* borkh)," *J. Food Sci.*, 59:581.

Eschleman, M., ed. 1984. *Introductory Nutrition and Diet Therapy*. London, Mexico City, New York, St. Louis, Sao Paulo, Sydney: J. B. Lippincott Co.

Feltman, J., ed. 1990. *Prevention's Giant Book of Health Facts*. Emmaus, PA: Rodale Press.

FAO/WHO. ed. 1994. *Summary of Evaluations Performed by the Joint FAO/WHO Expert Committee on Food Additives 1956–1993*. Geneva: Int. Life Sci. Inst. Press.

Gajek, O., M. Nabrzyski, and R. Gajewska. 1987. "Metallic Impurities in Imported Canned Fruit and Vegetables and Bee Honey," *Roczniki PZH*, 38:14 (in Polish).

Ha, Y. W., R. L. Thomas, L. A. Dyck, and M. E. Kunkel. 1989. "Calcium Binding of Two Microalgal Polysaccharides and Selected Industrial Hydrocolloids," *J. Food Sci.*, 54:1336.

Hendler, S. S., ed. 1990. *The Doctor's Vitamin and Mineral Encyclopedia*. New York, London, Toronto, Sydney, Tokyo, Singapore: Simon and Schuster.

Hultin, H. O. 1994. "Oxidation of Lipids in Seafoods," in *Seafoods; Chemistry, Processing Technology and Quality*, F. Shahidi and J. R. Botta, eds. London, Glasgow: Blackie Academic and Professional. Chapman and Hall, pp. 49–74.

Kraniak, J. M. and L. A. Shelef. 1988. "Effect of Ethylenediaminetetraacetic Acid (EDTA) and Metal Ions on Growth of *Staphylococcus aureus* 196 E in Culture Media," *J. Food Sci.*, 53:910.

Lopez, A., D. R. Ward, and H. L. Williams. 1983. "Essential Elements in Oysters (*Crassostrea virginica*) as Affected by Processing Method," *J. Food. Sci.*, 48:1680, 1961.

Marzec, Z., H. Kunachowicz, K. Iwanow, and U. Rutkowska, eds., 1992. *Tables of Trace Elements in Food Products*. Warsaw: National Food and Nutrition Institute (in Polish).

Miller, D. K., J. V. Gomez-Basauri, V. L. Smith, J. Kanner, and D. D. Miller. 1994. "Dietary Iron in Swine Affects Nonheme Iron and TBAR's in Pork Skeletal Muscles," *J. Food Sci.*, 59:747.

Miller, D. K., V. L. Smith, J. Kanner, D. D. Miller, and H. T. Lawless. 1994. "Lipid Oxidation and Warmed-Over Aroma in Cooked Ground Pork from Swine Fed Increasing Levels of Iron," *J. Food Sci.*, 59:751.

Nabrzyski, M. and R. Gajewska. 1984. "Determinations of Mercury, Cadmium and Lead in Food," *Roczniki PZH*, 35:1 (in Polish).

Nabrzyski, M. and R. Gajewska. 1995. "Aluminum and Fluoride in Hospital Daily Diets and Teas," *Z. Lebensm, Unters. Forsch.*, 201:307.

Nabrzyski, M., R. Gajewska, and A. Lebiedzińska. 1985. "Arsenic in Daily Food Rations of Adults and Children," *Roczniki PZH*, 36:113 (in Polish).

Oellingrath, I. M. and E. Slinde. 1988. "Sensory Evaluation of Rancidity and Off-Flavor in Frozen Stored Meat Loaves Fortified with Blood," *J. Food Sci.*, 53:967.

Pasławska, S. and M. Nabrzyski. 1975. "Assay of Iodine in Powdered Milk," *Bromat. Chem. Toxicol.*, 8:73 (in Polish).

Pearson, A. M., J. D. Love, and F. B. Shorland. 1977. " 'Warmed Over' Flavor in Meat, Poultry, and Fish," *Advances in Food Chemistry,* 23:2.

Ramanathan, L. and N. P. Das. 1993. "Effect of Natural Copper Chelating Compounds on the Prooxidant Activity of Ascorbic Acid in Steam-Cooked Ground Fish," *Internat. J. Food Sci. Technol.,* 28:279.

Rosenberg, J. H. and N. W. Solomons. 1984. "Physiological and Pathophysiological Mechanism in Mineral Absorption," in *Absorption and Malabsorption of Minerals, Vol. 12.* N. W. Solomons and J. H. Rosenberg, eds. New York: Alan R. Liss, Inc., pp. 1–13.

Rutkowska, U. 1975. "The Effect of the Grinding Process on Contents of Copper, Zinc, and Manganese in Rye and Wheat Flour," *Roczniki PZH,* 26:339 (in Polish).

Samat, S. K., R. S. Singhal, P. R. Kulkarni, and D. W. Rege. 1993. "Protein-Polysaccharide Interactions: A New Approach in Food Formulation," *Internat. J. Food Sci. Technol.,* 28:547.

Solomons, N. W. ed. 1984. *Absorption and Malabsorption of Mineral Nutrients.* New York: Alan R. Liss, Inc., pp. 269–295.

True, R. H., J. M. Hogan, J. Augustin, S. R. Johnson, C. Teitzel, R. B. Toma, and P. Or. 1979. "Changes in the Nutrient Composition of Potatoes during Home Preparation: III Minerals," *Amer. Potato J.,* 56:339.

Trzebska-Jeske, I., I. Nadolna, U. Rutkowska, and B. Secomska. 1973. "The Effect of Mechanical Processing on Nutritional Value of Groats Produced in Poland," *Roczniki PZH,* 24:717 (in Polish).

Vaessen, H. A. M. G. and A. van Ooik. 1989. "Speciation of Arsenic in Dutch Total Diets; Methodology and Results," *Z. Lebensm. Unters. Forsch.,* 189:232.

Wu, S. Y. and M. S. Brewer. 1994. "Soy Protein Isolate Antioxidant Effect on Lipid Peroxidation of Ground Beef and Microsomal Lipids," *J. Food Sci.,* 59:702.

WHO. Environmental Health Criteria, 18 ed. 1981. *Arsenic.* Geneva.

WHO. Environmental Health Criteria, 36 ed. 1984. *Fluorine and Fluorides.* Geneva.

WHO. Food Additives Series, 21 ed. 1986. *Toxicological Evaluation of Certain Food Additives and Contaminants.* Cambridge, 30th meeting of the Joint FAO/WHO Expert Committee on Food Additives.

WHO. Food Additives Series, 24 ed. 1989. *Toxicological Evaluation of Certain Food Additives and Contaminants.* Cambridge, 33rd meeting of the Joint FAO/WHO Expert Committee on Food Additives.

WHO. Food Additives Series, 30. ed. 1993. *Toxicological Evaluation of Certain Food Additives and Naturally Occurring Toxicants.* Geneva, 39th meeting of the Joint FAO/WHO Expert Committee on Food Additives.

WHO. Food Additives Series, 32. ed. 1993. *Toxicological Evaluation of Certain Food Additives and Contaminants.* Geneva, 41st meeting of the Joint FAO/WHO Expert Committee on Food Additives.

Williams, D. M. 1982. "Clinical Significance of Copper Deficiency and Toxicity in the World Population," in *Clinical Biochemical and Nutritional Aspects of Trace Elements, Vol. 6.* A. S. Prasad, ed. New York: Alan R. Liss, Inc., pp. 277–300.

Wilpinger, M., I. Schönsleben, and W. Pfanhauser. 1995. "Chrom in öesterreichischen Lebensmitteln," *Z. Lebensm. Unters. Forsch.* 201:421 (in German).

Wojnowski, W. 1994. "Skladniki mineralne." In *Chemiczne i funkcjonalne wlaściwości skladników żywności.* Z. E. Sikorski, ed. Warszawa: Wydawnictwa Naukowo-Techniczne, pp. 76–92 (in Polish).

CHAPTER 4

Saccharides

PIOTR TOMASIK

NATURAL FOOD CARBOHYDRATES, OCCURRENCE, ROLE, AND APPLICATIONS

NATURE commonly utilizes carbohydrates as a source of energy, structure-forming material, water maintaining hydrocolloids and even sex attractants. All organism cells contain saccharide components in their membranes. Frequently, saccharides exist in naturally derivatized forms, e.g., aminated such as chitin and chitosan, esterified, alkylated, oxidized, reduced, or linked to proteins, lipids, and other structures such as hemoglobin, hesperidin, etc.

Lower monosaccharides, i.e., aldo- and keto-bioses, trioses, and tetroses, do not exist in a free state. Glyceraldehyde and hydroxyacetone in phosphorylated forms are the products of alcoholic fermentation and glycolytic sequence. Erythrose and erythrulose also appear in phosphorylated forms in the pentose cycle of glucose, while ketopentose-ribulose can be found as its phosphate ester (Table 4.1).

CARBOHYDRATE STRUCTURE

Carbohydrates are either polyhydroxyaldehydes (aldoses and oses) or polyhydroxyketones (ketoses and uloses); there is an electron gap at their carbonyl carbon atom. Thus, they accept nucleophiles to form hydrates with water or hemiacetals, acetals and hemiketals, respectively, with alcohols. In pentoses, pentuloses, hexoses, hexuloses, and higher carbohydrates, one of the hydroxylic groups can play the role of internal

65

TABLE 4.1. Natural Saccharides, Their Occurrence, and Application.

Saccharide	Occurrence	General Applications
Monosaccharides		
Aldopentoses nonderivatized		
L-Arabinose	Plant gums, hemicelluloses, saponins, protopectin	Alcoholic fermentation, source of furfural, culture medium for certain bacteria
D-Xylose	Accompanies L-arabinose, xylanes, hemicelluloses, plant gums, glycosides, saponins	Reduction to xylitol, substitute of sucrose, alcoholic fermentation, source of furyl-2-aldehyde, furfural
Aldohexoses nonderivatized		
D-Galactose	Widespread in oligo- and polysaccharides, plant mucus and gums, glycolsides, saponins	Diagnostics in liver tests
D-Glucose	Widespread in plants and animals, honey, component of inverted sugar, saponins	Alcoholic fermentation, sweetener, energy pharmacopeal material, yeast nutrition, food preservative
Ketohexoses nonderivatized (Hexuloses)		
D-Fructose	Fruits, honey, inverted sugar	Non-cavity-causing sweetener, sweetener for diabetics, food preservative, liver detoxication
L-Sorbose	Rowan berries	Source for ascorbic acid synthesis
Ketohexoses derivatized		
D-Glucosamine	Chitin and chitosan	Pharmaceutical aid, antiarthritic drug
Disaccharides		
Sucrose	Sugar beet, sugar cane, widespread in other plants	Alcoholic fermentation, sweetener, caramel, food preservative

(continued)

TABLE 4.1. (continued).

Saccharide	Occurrence	General Applications
Polysaccharides		
Agar	Red algae	Nutritive media in microbiology, gel-forming agent, emulsifier, bread staling retardant, meat texturizer, and meat substitute
Alginates	Brown algae cell membranes	Thickener, gel-forming agent, food and beer foam stabilizer
Carrageenans:	Red sea weeds	Gel-forming agent, stabilizer, protein fiber texturizer, fat milk anticoagulant, milk clarifying agent
ɩ-Carrageenan		
ϰ-Carrageenan		
γ-Carrageenan		
μ-Carrageenan		
ν-Carrageenan		
Cellulose	Widespread in plants	Saccharification to glucose, ballast food component, chromatographic sorbent
Dextran	Frozen sugar beets	Chromatographic sorbent (Sephadex), blood substitute
Furcellaran	Red seaweed	Gel-forming food additive, filler, marmalade stabilizer, protein precipitant
Gatti gum	Anageissus latifolia tree	Emulsifier, stabilizer
Guaran gum	Leguminous plant Cyanopsis tetragonoloba seeds	Food, cosmetic, and pharmaceutical thickener and stabilizer
Gum Arabic	Acacia (Senegal) tree	Emulsifier, stabilizer in baked products, flavor fixative
Gum karaya	Sterculiacea tree (India)	Foam stabilizer, thickener

(continued)

TABLE 4.1. (continued).

Saccharide	Occurrence	General Applications
Gum tragacanth: Tragacanthin Bassarin	Astragalus species (Middle East)	Thickener, stabilizer
Glycogen	Liver, muscle	Glucose reservoir
Hemicelluloses: Arabinogalactan Galactan Mannans Xylans	Larch Widely spread in plants	Ballast food components Emulsifier, stabilizer Alcoholic fermentation, reduction to alcohols
Heparin	Liver, tongues	Blood anticlotting agent
Hialuronic acid	Connective tissues	Water binder
Inulin	Plant roots of *Compositae* and *Campanulaceae*	Glucose reservoir
Locust bean gum	Locust bean	Thickener, binder, stabilizer
Pectin	Widely distributed in plants, mainly in apples and citrus fruits	Gel-forming agent, beer stabilizer
Protopectin	Widely distributed in plants	Decomposes to pectin on fruit maturation and cooking
Starch: Amylose Amylopectin	Widely distributed in plants	Food filler, thickener, gel-forming agent, bakery products; saccharification to syrups
Tamarind flour	Tamarind tree (India)	Thickener, marmalade, jelly, ice cream, and mayonnaise stabilizer
Xanthan gum	*Xanthomonas camprestris* bacteria cells	Hydrocolloid stabilizer

nucleophile. Thus, internal hemiacetals and ketals are formed, respectively, all with either five- or six-membered cycles.

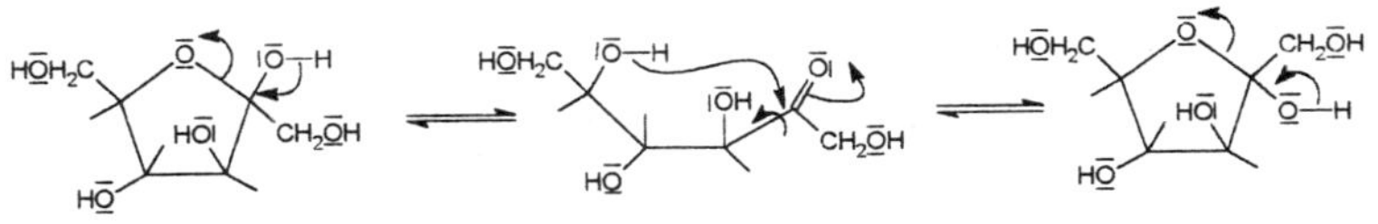

(4.1) (4.2) (4.3)

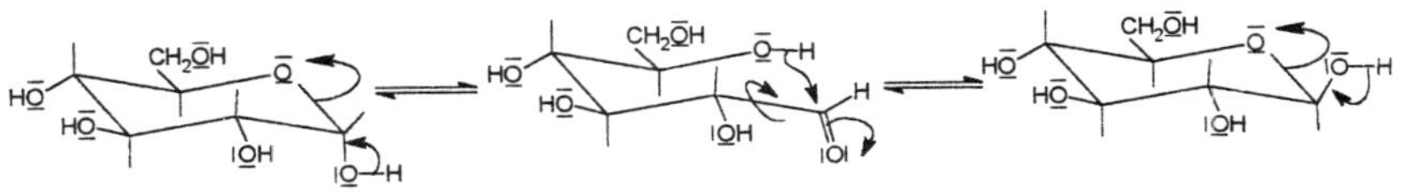

(4.4) (4.5) (4.6)

Any cyclization of open-chain compounds into either five- or six-membered rings, if sterically possible, is always energetically beneficial. Such rings are thermodynamically stable.

Since all carbon atoms in the cycles are sp^3-hybridized, the bond angles of about $109°$ have to be retained. It is possible only if the cycles take nonplanar conformations. The low symmetry of the rings allows the strainless conformations.

In some cases, the pyranose ring formation can either be obstructed or blocked and a five-membered furanose ring dominates for the given sugar. The molecular structure of di- and higher saccharides is additionally controlled by a potential energy benefit resulting from the formation of intramolecular hydrogen bonds, as in cellobiose, lactose, sucrose, and maltose.

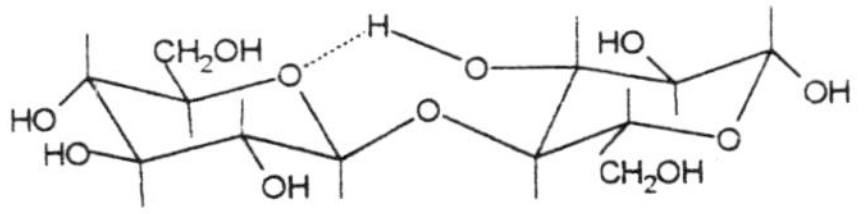

Cellobiose

(4.7)

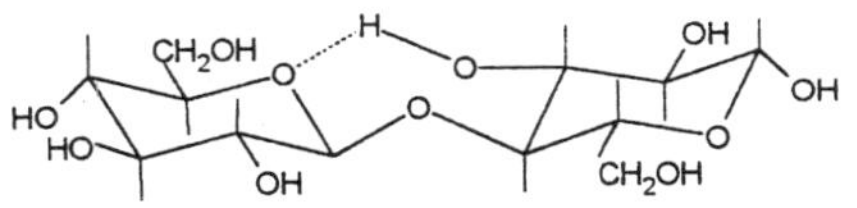

Lactose

(4.8)

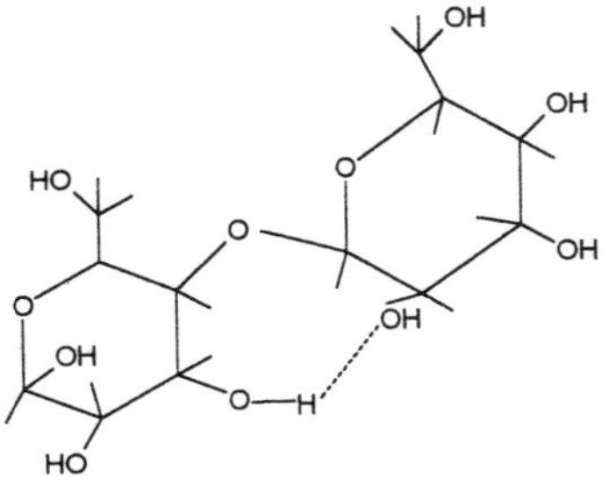

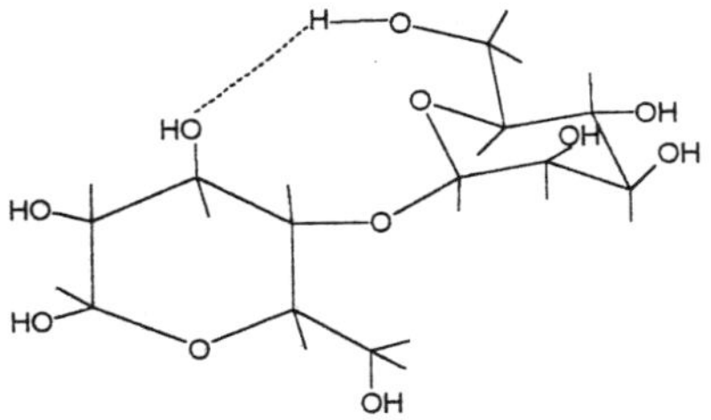

Maltose

(4.9)   (4.10)

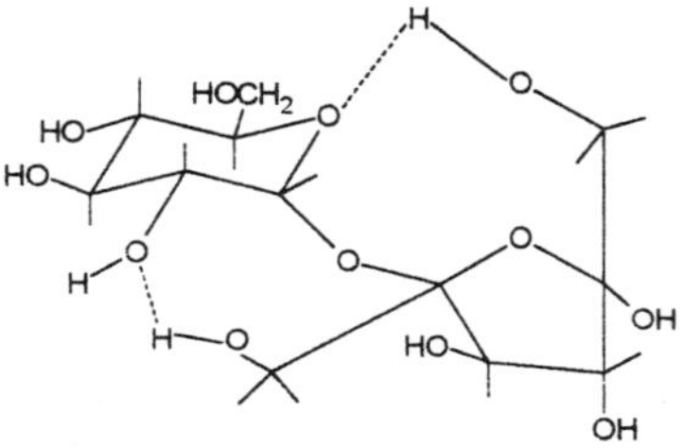

Sucrose

(4.11)

Both maltose structures correspond to two energy minima. Two hydrogen bonds stabilize the sucrose structure. This is possible only if the fructose moiety takes the furanosyl structure.

In polysaccharides, the structural elements are even more important. In a single polysaccharide chain, the regularity of its structure, i.e., number of different saccharide units in the chain, branching of the chain, and the presence of either more polar groups (COOH, PO_3H_2, SO_3H) or more nonpolar groups (OMe, NAc) than the OH group have to be considered as crucial factors for the overall structure. Amylose and cellulose are the most regularly built. They form polymer chains of α-D- and β-D-glucose units, respectively. In very random cases, amylose chain is branched with short chains. The amylose chain is originally a long, randomly coiled structure that turns into more ordered forms, depending on the interaction-type offered by the environment. Thus, alcohols, lipids, and fatty acids cause coiling of the amylose chain. The helix cavity is hydrophobic, and the majority of the hydroxy groups are situated on the external surface of the helix (the V-type amylose). The number of glucose units in the helix turn depends on the agent responsible for such coiling. With KOH, one turn takes six glucose units, whereas KBr reduces the turn to four glucose units; *tert*-butanol expands the turn into seven glucose units and

α-naphthol to eight glucose units. An additional stabilization of the amylose helix comes from the double helix formation. Depending on the helix–helix interactions and in consequence of their mutual arrangement, A- and B-type amylose is formed. Recently distinguished C-type amylose appears to be the combination of both A- and B-patterns.

The β-D-glucose unit conformation in cellulose offers a particularly strong hydrogen bond cross-linked macrostructure of this polysaccharide. Glycans with 1,2- 1,3-, and particularly, 1,6-linked units have a more irregular, loosely jointed structure.

The heterogeneity of the polysaccharide structural units and their volumes introduce further irregularities in the macrostructure. Similar effects can sometimes be noted when the polarity of the hydroxy groups is reduced by their alkylation. An opposite effect can be achieved in the presence of more polar groups. Their charge is an additional ordering factor. Amylopectin presents a special case. It is a highly branched (Figure 4.1) homopolymer of α-D-glucose units. The $1 \rightarrow 6$ linked branches, which occur at about each eighth glucose unit, contain fifteen to thirty glucose units. These terminal branches behave like amylose; i.e., they are capable of coiling at appropriate conditions. In spite of an irregular, bulky structure, amylopectin also forms a double helix.

Because of functional properties, the structure of the polysaccharide matrix, the tertiary structure, is also essential. In solution, polysaccharides such as amylose form separate fibrils that, upon cooling, turn either into micelles at a low-temperature gradient or into gels at a high-temperature gradient. Amylose and amylopectin coexist in the starch granule organization. This semi-native mixture has specific functional properties resulting from a different amylose-to-amylopectin ratio; size of granules (Table 4.2); content of residual components of native starch, e.g., lipids, proteins, cations; and random esterification with phosphoric acid.

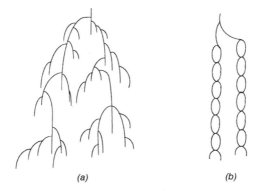

(a)   *(b)*

Figure 4.1 Scheme of (a) amylopectin molecule and (b) its terminal fragment.

TABLE 4.2. **Properties of Starch of Various Origin.**

Origin	Amylose Content (%)	Granule Size (μm)	Gelation Temperature (°C)
Barley	19–22	5–40	51–59
Maize	21–24	10–30	67–100
Oat	23–30	5–15	87–90
Potato	18–23	1–100	59–68
Rice	8–37	2–10	68–78
Rye	24–30	8–60	55–70
Triticale	23–24	2–40	55–62
Waxy maize	1–2	10–30	62–72
Wheat	24–29	2–38	59–64

The gelation temperature corresponds to the opening of the granule shell and migration of its interior into solution to form gel.

Cellulose that is fully insoluble forms microfibrils that are composed of crystallites and amorphous regions. Such amorphous regions are also met within starch granules. Perhaps they are areas occupied by native amylose.

CARBOHYDRATE CHIRALITY

All carbohydrates, including polysaccharides, are chiral. In the majority of mono- and oligosaccharides, the chirality expressed as specific rotation, $[\alpha]^0$, is a concentration independent magnitude

$$[\alpha]^0 = (\alpha \cdot 100)/lc \qquad (4.1)$$

where:

α = angle twist determined in the polarimeter
l = length of a polarimetric tube
c = concentration in g/100 cm^3

It varies with time and stabilizes with equilibrium. A reducibility of a given saccharide is a necessary condition for the chirality variation called mutarotation. Aqueous D-glucose at 25°C has approximately 0.003% of the open-chain compound in which a free rotation occurs around the C−C single bond.

The mutarotation of oligosaccharides may proceed stepwise due to independent mutarotation of all monosaccharide units, provided they have a free hydroxy group at the anomeric carbon atom.

CARBOHYDRATE REACTIVITY

CHEMICAL AND PHYSICAL TRANSFORMATIONS OF MONO-, DI-, AND OLIGOSACCHARIDES

Reactions of Aldehyde and Ketone Functions

Mutarotation

Mutarotation is either base or acid catalyzed. It has a limited, rather diagnostic, significance in food chemistry and technology. Practical use of this reaction is demonstrated in milk powder manufacture. Evaporation of milk at a rate lower than mutarotation of lactose gives a product containing less α-lactose isomer, which crystallizes in prism- or pyramid-like form. A rapid milk evaporation gives an amorphous mixture of α- and β-lactose.

Reduction to Alcohols

The reduction carried out on an industrial scale involves either $NaBH_4$ or electrochemical or catalytic hydrogenation. The resulting sugar alcohols are utilized as sweeteners for diabetics and in food canning. Because of their hygroscopicity, they are used as food humectants. The reduction leads to open-chain polyols with a new chiral center. In consequence, each ketose gives two alcohols, whereas each aldose gives only one alcohol.

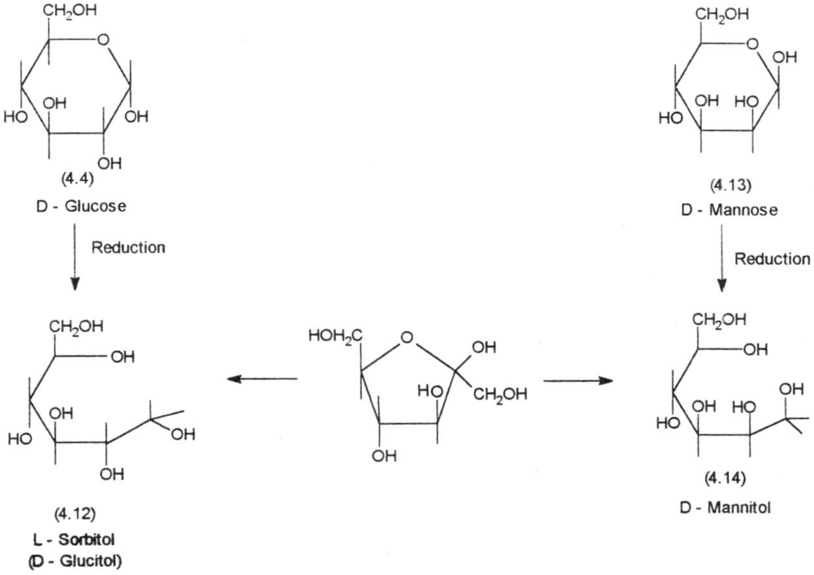

(4.4)
D - Glucose

(4.13)
D - Mannose

Reduction

Reduction

(4.12)
L - Sorbitol
(D - Glucitol)

(4.14)
D - Mannitol

Addition to the Carbonyl Group

The internal, cyclic hemiacetal formation is one of the illustrations of such an addition. The H_2N-X nucleophiles, with X being NH_2 (hydrazine), NHAr (arylhydrazines), OH (hydroxylamine), $NHCONH_2$ (semicarbazide), $NHCSNH_2$ (thiosemicarbazide), or alkyl (primary amine), give the following products:

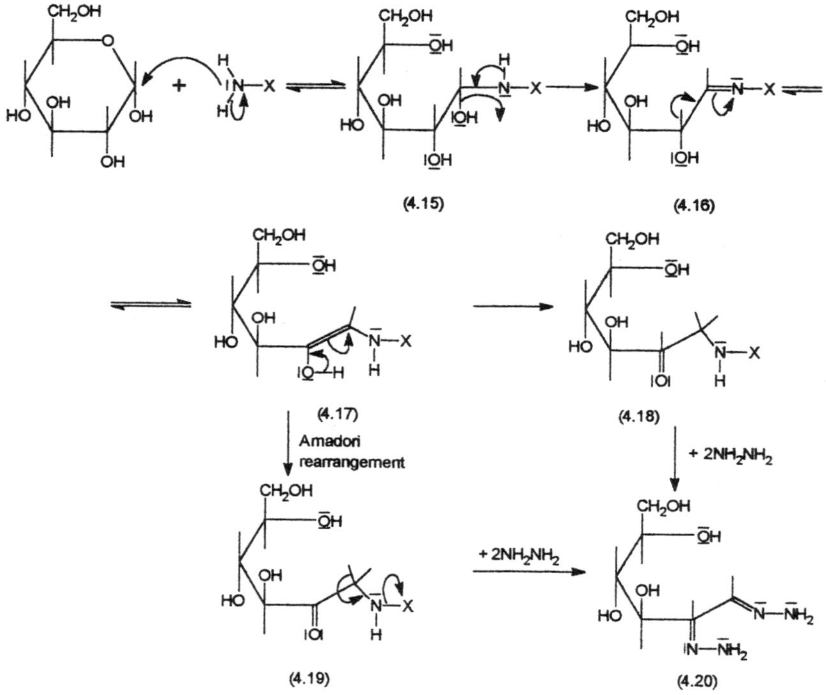

(4.15)　　　(4.16)

(4.17)　　　(4.18)

Amadori
rearrangement

(4.19)　　　(4.20)

X = NH_2−hydrazone; NAHr−arylhydrazone; OH−oxime; $NHCONH_2$−semicarbazone; $NHCSNH_2$−thiosemicarbazone; alkyl−imines (Schiff bases). The reaction with either hydrazine or arylhydrazine proceeds further to give osazones. Ketoses react similarly with involvement of the vicinal CH_2OH group. These reactions are typical for saccharides, α-hydroxyaldehydes and α-hydroxyketones. The reactions with mono-, di-, and lower oligosaccharides have solely some analytical value. These compounds can readily be purified by crystallization. They have characteristically sharp melting points and crystalline forms specific for a given sugar derivative. The reactions with amino acids, nucleotides, proteins, or ammonia are important in food chemistry. Aldosylamines undergo the Amadori rearrangement and subsequently turn into caramels (the Maillard reaction) and/or heteroaromatic compounds—derivatives of pyrrole, imidazole, and pyrazine constituting so-called secondary food aromas. Ketoses react similarly into ketosylamines, which in the first step, undergo the Heyns rearrangement.

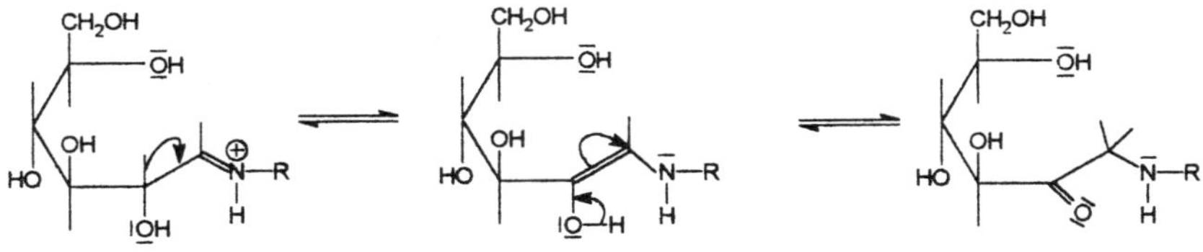

(4.21)   (4.22)   (4.23)

Oxidation

The oxidation of aldoses with bromine in alkaline solution leads to aldonic acids that readily self-esterify (lactonize) into δ- and γ-lactones. Lactones and free acids exist in an equilibrium as is shown for the oxidation of D-glucose to D-gluconic acid and both glucono lactones. β-Conformers oxidize more readily than α-conformers

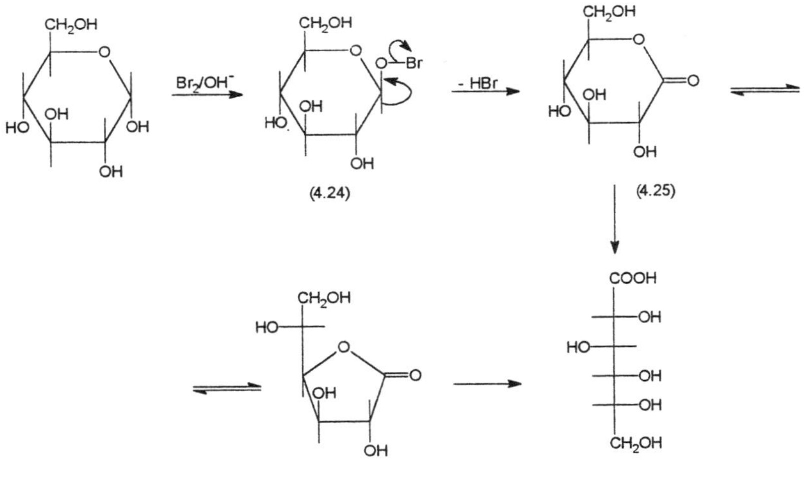

(4.24)   (4.25)

(4.26)   (4.27)
D - Gluconic acid

Glucono-δ-lactone has found its application in baking powders, raw fermented sausages, and dairy products where the slow release of acid is required.

The oxidations with Cu^{2+} and Bi^{3+} ions serve only as the analytical tests for reducing sugars.

Polymerization and Reversion

Such reactions proceed in strongly acidic media. This is the addition of subsequent saccharide nucleophiles to the electron gaps at the anomeric carbon atoms.

Reactions of the Hydroxylic Groups

Esterification

Saccharides are commonly esterified with acyl chlorides, as well as organic and inorganic anhydrides. Esterification, particularly with organic acyl groups, can be either exhaustive or selective. Selective acylation has to proceed under protection of certain hydroxylic groups. All hydroxylic groups of D-glucose except that at C3 can be protected by the reaction with acetone in acidic medium. 1,2,5,6-Di-*O*-isopropylidene-α-D-glucofuranose is subsequently acylated with anhydride followed by decomposition of monoacylated diketal with carboxylic acid.

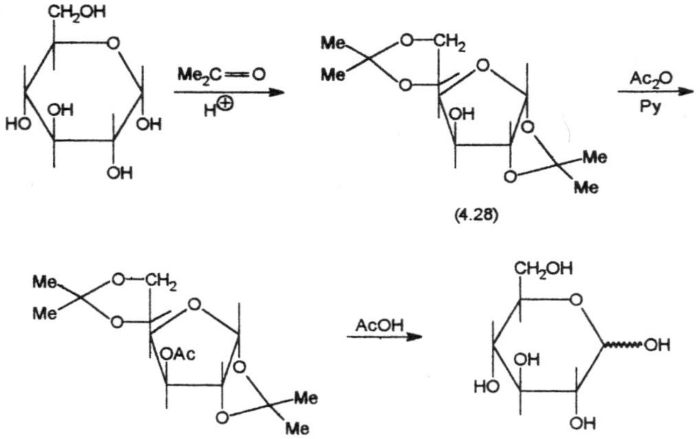

(4.28)

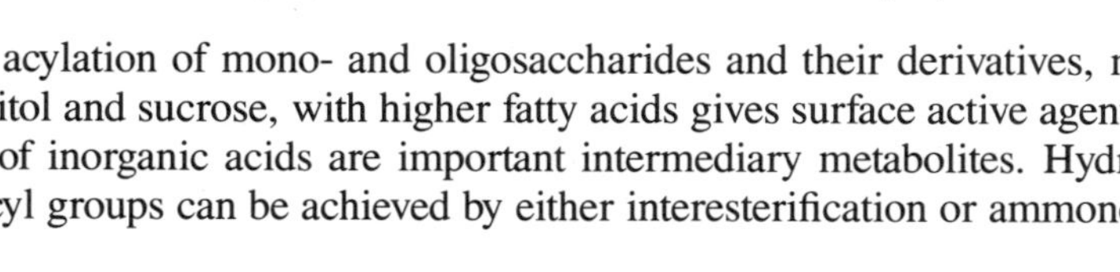

(4.29)   (4.30)

The acylation of mono- and oligosaccharides and their derivatives, mainly sorbitol and sucrose, with higher fatty acids gives surface active agents. Esters of inorganic acids are important intermediary metabolites. Hydrolysis of acyl groups can be achieved by either interesterification or ammonolysis.

Etherification

There are several methods of sugar etherification, e.g., (1) the Fischer glycosidation with alcohols catalyzed by hydrochloride; (2) the Haworth methylation with dimethyl sulfate in an alkaline medium, which is an exhaustive methylation with configuration retention at the anomeric carbon atom; (3) the Irvine-Purdie methylation, which involves methyl iodide in the presence of Ag_2O, which is exhaustive methylation; and (4) the Koenigs-Knorr glycosidation, which is the substitution of an α-halo atom of α-halopentacetyl sugars with methanol in the presence of Ag_2CO_3. Such reactions with small alkyl groups are important for saccharide structural analysis.

Halogenation

The hydroxyl group at C1 is readily substituted with a halide atom (X) when a pentacetylated sugar is treated with HX in acetic acid. Exhaustive replacement of all hydroxyl groups can be achieved with reagents suitable for simple alcohols, i.e., $COCl_2$, $SOCl_2$, $POCl_3$, PCl_3, and PCl_5.

Dehydration

This is the first step of sugar caramelization. It is an intramolecular elimination of one water molecule producing 1,6-anhydrosugar or epoxy product.

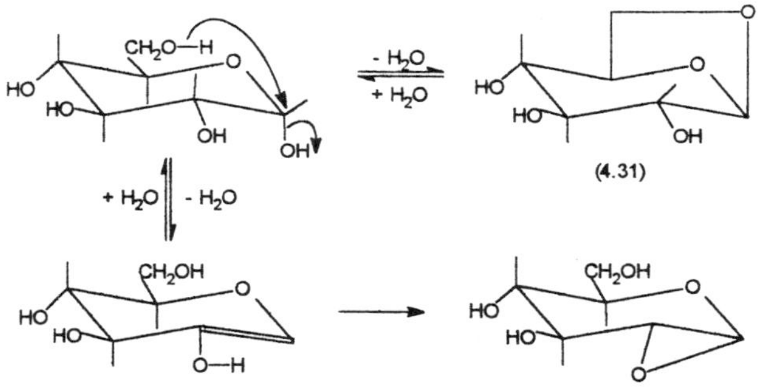

(4.31)

(4.32)   (4.33)

Other anhydrosugars are 3,6-anhydroglucose (a) and 2,5-anhydromannose (chitose) (b).

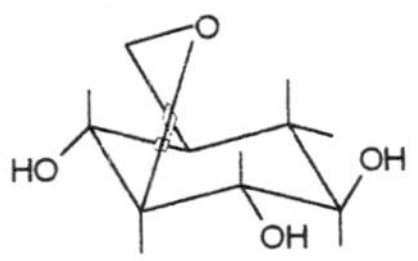

(4.34) 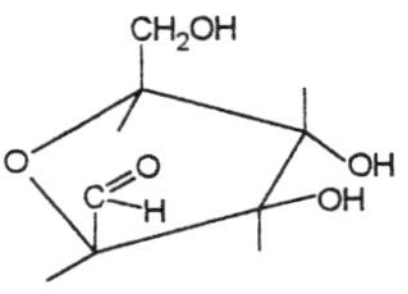   (4.35)

Further heating of dehydrated sugars results in the formation of three subsequent compounds called caramelan, caramelen, and caramelin.

$$6C_{12}H_{22}O_{11} - 12H_2O = 6C_{12}H_{12}O_9 \text{ (caramelan)}$$

$$6C_{12}H_{22}O_{11} - 18H_2O = 2C_{36}H_{18}O_{24} \text{ (caramelen)}$$

$$6C_{12}H_{22}O_{11} - 27H_2O = 3C_{24}H_{26}O_{13} \text{ (caramelin)}$$

Reduction

A multistep reaction leads to desoxysaccharides. It involves 1-halopentacetylated saccharide, which is dehalogenated with zinc into acetylated glucal. Hydrolyzed glucal accepts sulfuric acid into its double bond to give 2-desoxysaccharide.

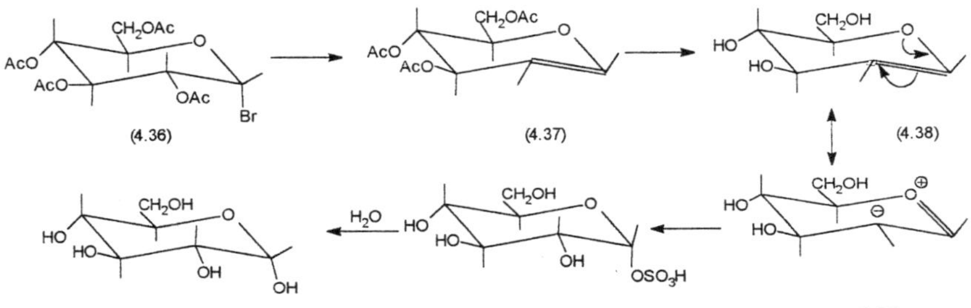

(4.36) (4.37) (4.38) (4.39) (4.40)

Oxidation

Apart from CO_2 and H_2O, there are three series of products that result from the oxidation of saccharides. They are 2,3-dialdehydes formed on the oxidative cleavage of saccharides with periodates. Further oxidation of dialdehydes leads to glyceric acid, glyoxalic acid, and corresponding carboxylic acids, as shown below for the oxidative cleavage of sucrose and maltose.

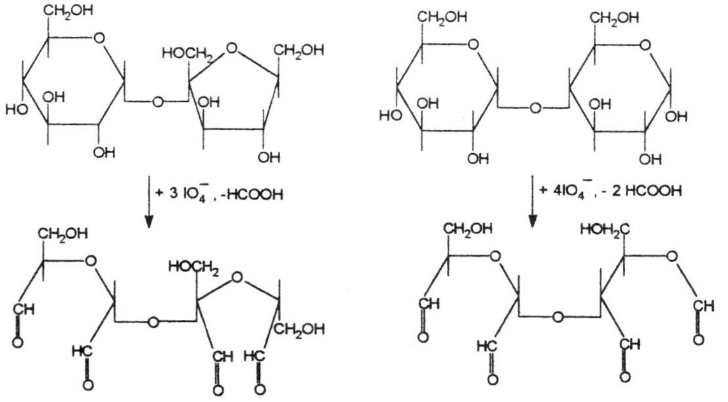

(4.41) (4.42)

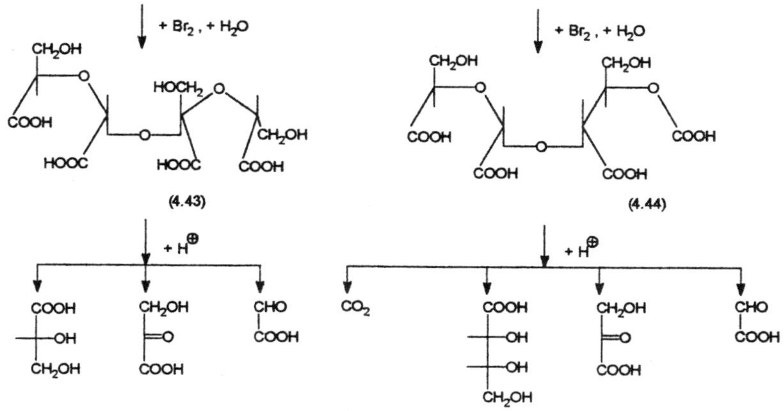

(4.43)

(4.44)

(4.45) D - Glyceric acid

(4.46) Hydroxypyruvic acid

(4.47) Glyoxalic acid

(4.48) Erythronic acid

The oxidation of monosaccharides with strong oxidants proceeds at both C1 and C6 atoms, leaving dicarboxylic, aldaric acids. Aldaric acids can lactonize. Lactones produce uronic acids when reduced with sodium amalgam.

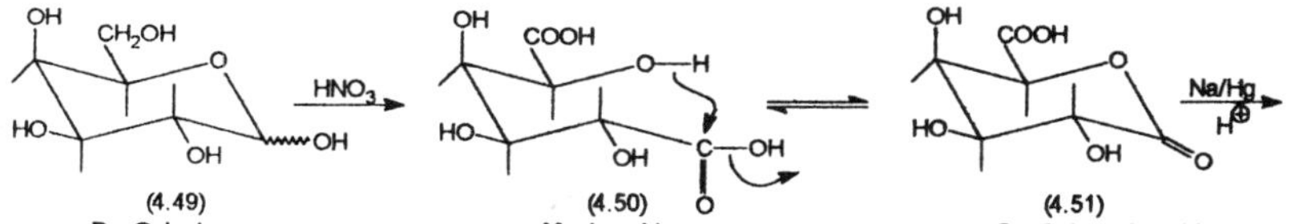

(4.49) D - Galactose

(4.50) Mucic acid

(4.51) D - Galactaric acid monolactone

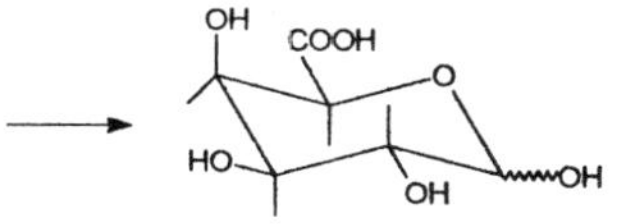

(4.52) D - Galacturonic acid

The oxidation at C1 can be prevented by protection of the 1-OH group. The glycosidic bond in oligosaccharides offers sufficient protection.

Complex Formation

The hydroxyl groups offer two types of interactions with molecules having either clearly dipol character or charge, i.e., ions. The hydrogen atoms of these groups are capable of interactions with electron excessive sites of dipoles and anions, whereas line electron pairs of the oxygen atom are electron donors for cations and the positive side of dipolar molecules.

Complexes of saccharides with metal salts were studied for more than 120 years. The complex formation is a general ability of saccharides and metals. Fruitful results were noted in the case of Ca^{2+} salts, preferably chloride and carbonate. The cation forms fairly stable compounds. This property was widely utilized in sugar manufacture for the separation of sugar from its syrups. Saccharide alcohols also coordinate metal ions. Depending on cations, their optical rotation is affected to a different extent but always in the order of $Na^+ < Mg^{2+} < Zn^{2+}$, $Ba^{2+} < Sr^{2+} < Ca^{2+}$. Complexes of saccharides with the cations of lanthanides, Mo, V, Cr, Mn, Fe, Co, Ni, Cu, Zn, Pb, U, Ag, As, Si, Sn, and Nb were also prepared and characterized. The complexes are weak. Recently, complexation of several mono-, di-, and trisaccharides with starch were characterized. The energies of the complex formation are low and range between 2 and 4 kJ/mol in the case of all D-glucose, D-galactose, D-fructose, D-mannose, D-lactose, D-maltose, and D-xylose. Such complexation has an impact to food rheology and texture. In spite of low energy, the complexation itself can seriously affect the sugar metabolism and biotechnological processes.

Reactions of Glycosidic Bond

As a typical acetal bond, a glycosidic bond readily hydrolyzes in an acid-catalyzed reaction. In this manner, di- and oligosaccharides can be split into monosaccharides. It is a common method for manufacturing of invert sugar, a mixture of α-D-glucose and D-fructose, from sucrose.

Specific Reactions of Saccharides

In strongly acidic media, saccharides produce furan derivatives in a sequence of reactions that are rearrangements and dehydrations followed by cyclization. Pentoses give furan-2-aldehyde and hexoses give 5-hydroxymethylfuran-2-aldehyde. This is also a pathway to maltol, 3-hydroxy-2-methyl-pyran-4-one and isomaltol, 2-acetyl-3-hydroxyfuran, both being responsible for baked bread aroma.

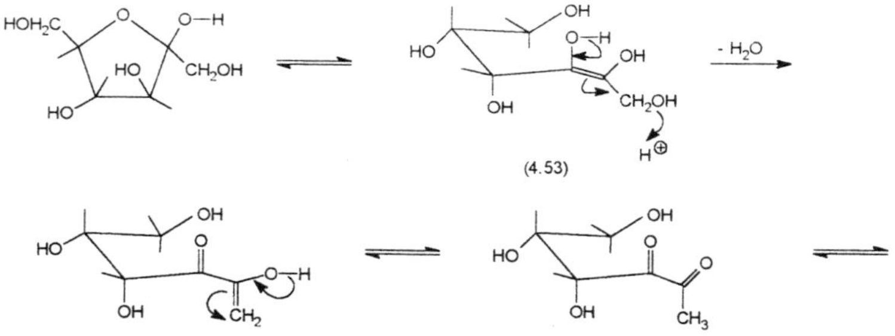

(4.53)

(4.54)   (4.55)

1 - Desoxy - D - erythro - 2,3 - hexadiulose

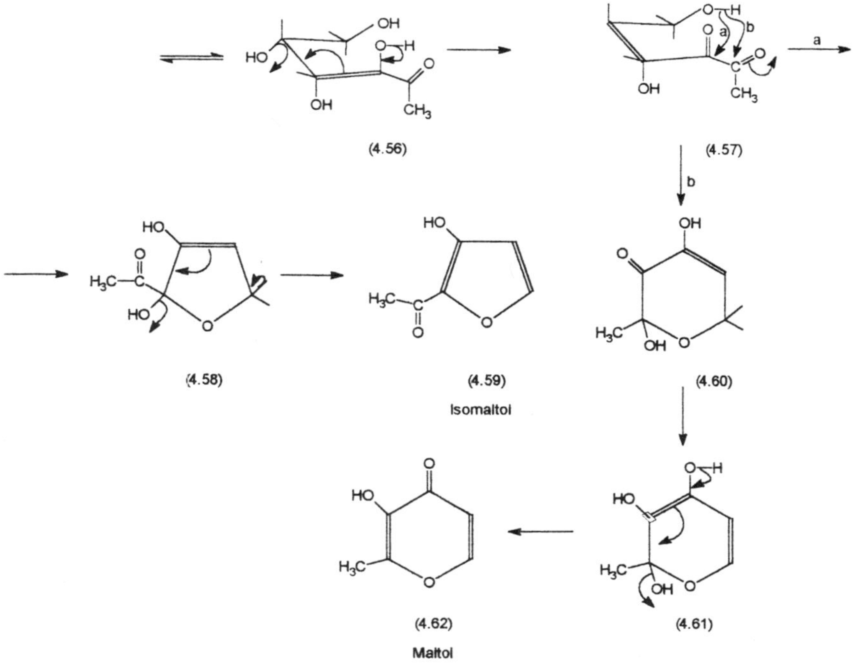

(4.56)   (4.57)

(4.58)   (4.59)   (4.60)
           Isomaltol

(4.62)   (4.61)
Maltol

In weakly acidic media, the reactions proceed at a lower rate. Reductones are formed with carbonyl group vicinal to an endiol moiety. The reductones are stable at pH < 6 and act as natural antioxidants. In strongly alkaline media, aldoses and ketoses readily enolize.

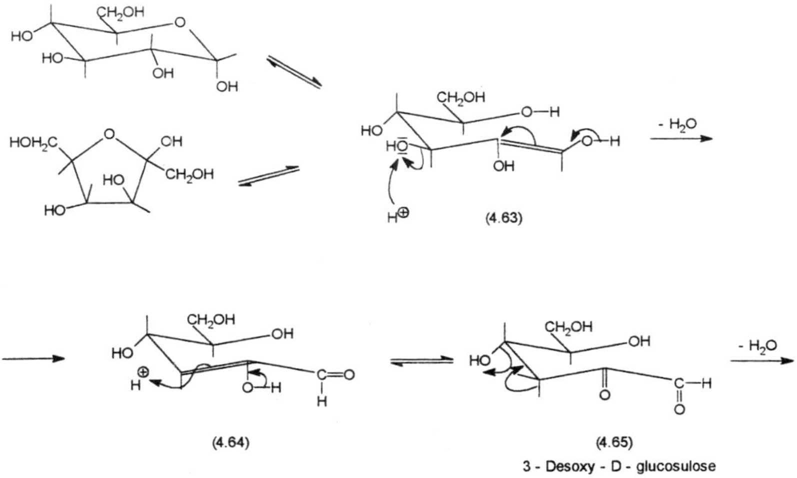

(4.63)

(4.64)   (4.65)
           3 - Desoxy - D - glucosulose

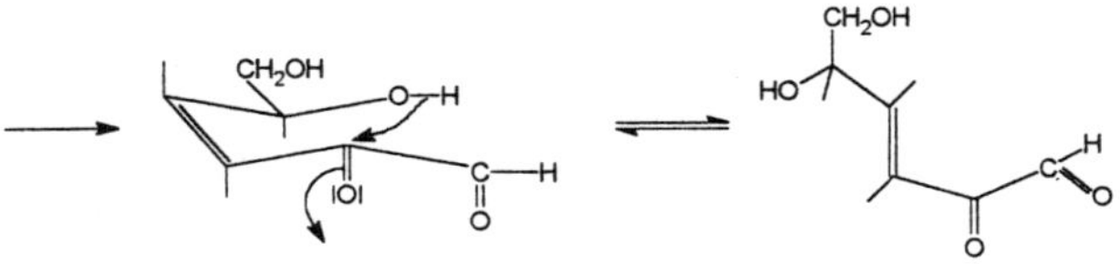

(4.66)   (4.67)

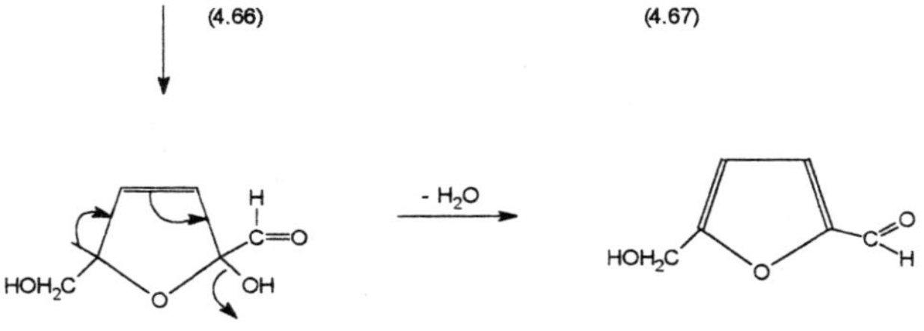

(4.68)   (4.69)

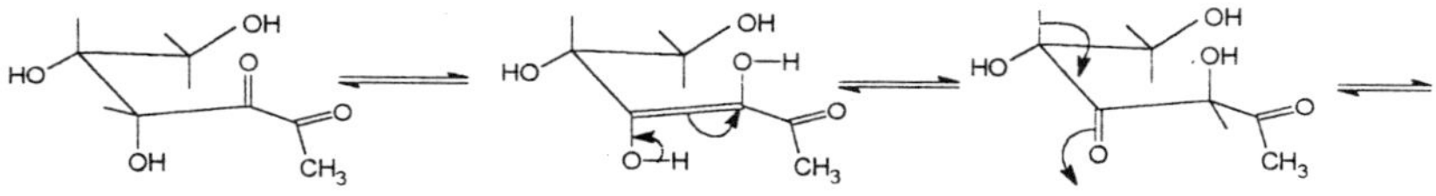

(4.70)   (4.71)

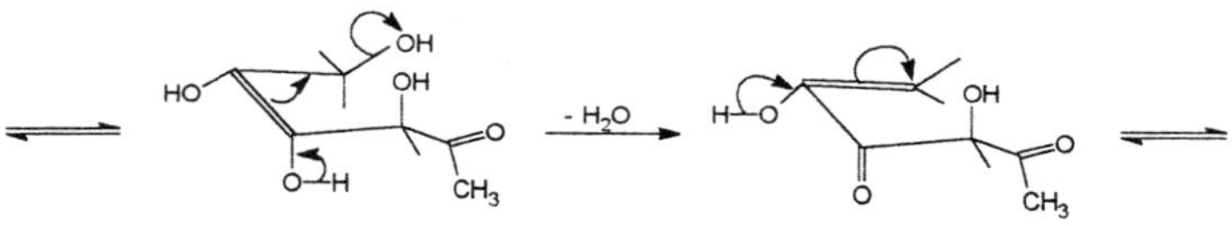

(4.72)   (4.73)

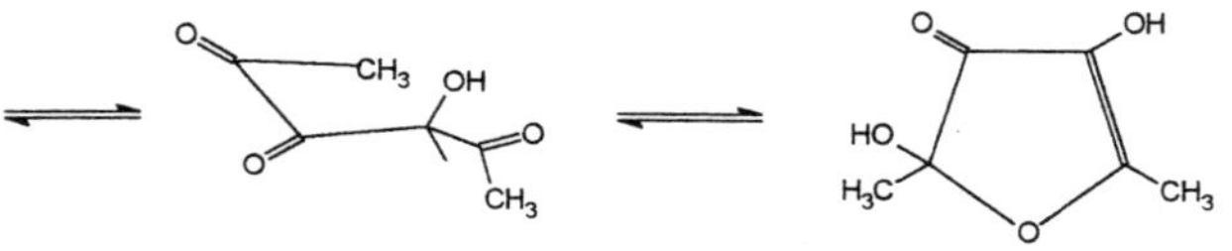

(4.74)   (4.75)

Diacetylformosine

Endiols can isomerize into other saccharides in the Lobry de Bruyn–van Ekenstein rearrangement. Thus, glucose can isomerize into mannose and fructose, accompanied by a small amount of psicose. Alkaline medium provides isomerization of disaccharides, which turn from aldoses into ketoses, as shown for lactose isomerized to lactulose.

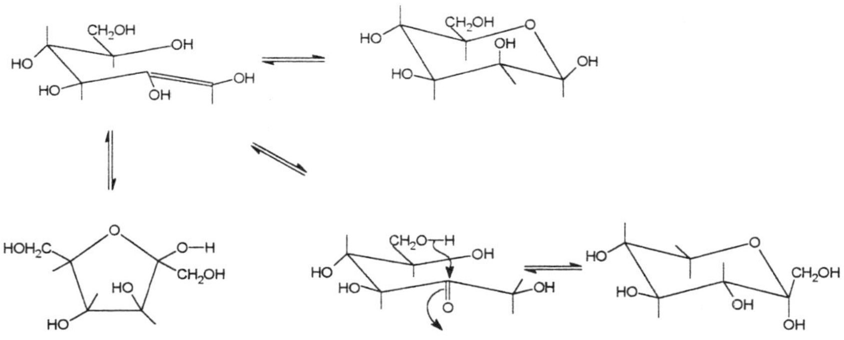

(4.76)

(4.77)

D - Psicose

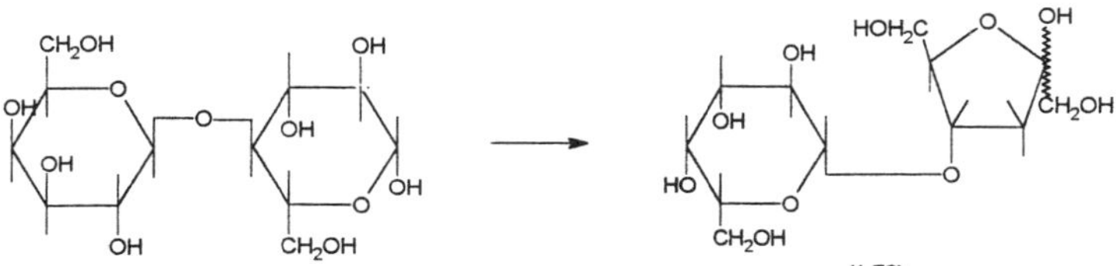

(4.78)

Lactulose

With either atmospheric oxygen or another oxidant, a partial oxidative degradation takes place.

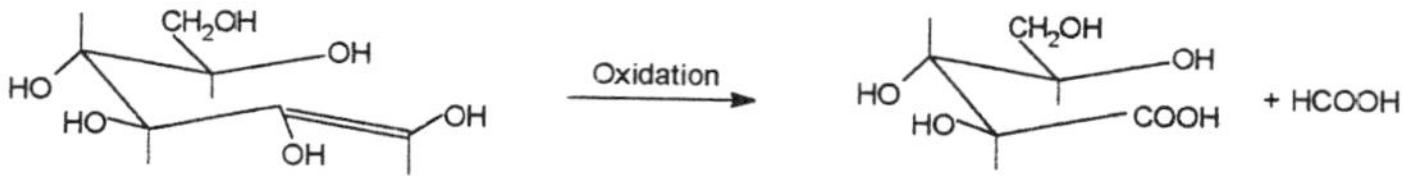

(4.79)

Since the enolization is not restricted to the 2- and 3-positions, a number of products are formed that undergo subsequent aldol condensations and the Cannizzaro oxidation. They are all 2-hydroxy-3-methyl-, 3,4-dimethyl-2-hydroxy, 3,5-dimethyl-2-hydroxy, and 3-ethyl-2-hydroxy-2-cyclopenten-1-ones, sugar acids, acetic acid, hydroxyacetone, three isomeric hydroxy-2-butanones, γ-butyrolactone, and such furan derivatives as furfuryl alcohol, 5-methyl-2-furfuryl alcohol, and 2,5-dimethyl-4-hydroxy-3-(2H)-furanone. They are food flavoring agents.

CHEMICAL AND PHYSICAL TRANSFORMATIONS OF POLYSACCHARIDES

Only a few among the known polysaccharides, e.g., starch, cellulose, and hemicelluloses, are modified in industrial scale. Products of their modifications are widely utilized in food chemistry. Even such food processing as cooking, baking, frying, and pickling frequently deal with food carbohydrate transformations.

The functional group reactivity in polysaccharides is, to a great extent, obstructed by their macrostructure. Only these reaction sites that reside on the surface of macrostructures become available for many reagents. Even so, many reactions are governed by heterogeneity of the reaction system. Only randomly reacting polysaccharides are solvent soluble. Among polysaccharides utilized either on or after transformation, hemicelluloses are exclusive in this respect. They are either water-soluble or, at least, they swell. Problems of the solubility and compactness are particularly clear in polysaccharides with their compact structure, which is fibrillar in cellulose and granular in starch. The reactivity in terms of the rate and degree of transformation can be controlled either by application of penetrating reagents, by loosening of the compact structure of the polysaccharide, or by involving physical action, e.g., the solvent effect, high pressure, sonification with ultrasound, swelling in alkaline solution, and temperature.

Depolymerization of Carbohydrates

If not utilized in the pulp industry, hemicelluloses used to be hydrolyzed in the acid-catalyzed processes mainly to monosaccharides and to furyl-2-aldehyde (pentosanes) and 5-hydroxymethyl-2-aldehyde (hexosanes). Monosaccharide-containing syrups, after purification, are either fermented or utilized as wood molasses for feeding ruminants. In another approach, xylose, the least soluble component of the syrup, is allowed to crystallize. Separated xylose is then hydrogenated over a Ni/Al catalyst at 120°C under 6×10^6 Pa into xylitol, a sweetener for diabetics, chewing gum, and toothpastes. Monoanhydrate of xylitol is utilized as a glycerol substitute. Hemicelluloses, together with proteins, are capable of the Maillard reaction. Thus, they contribute to the overall secondary aroma of processed foodstuffs.

The acid catalyzed hydrolysis of cellulose, the saccharification, results in splitting the terminal glucose unit of the fibers. Thus, under thorough control of reaction conditions, D-glucose is the sole product. A small amount of contamination results from unavoidable acid-catalyzed decompositon of D-glucose. Glucose syrup is either a source of pharmaceutical grade D-glucose or it is fermented to ethanol.

β-Glycosidic bonds of cellulose can be split thermally. Perhaps the most ancient polysaccharide processing—the dry wood distillation—delivers charcoal, water, tar, methanol, acetone, acetic acid, and gases. Liquid and gaseous fractions result from the thermolysis of thermally split D-glucose.

Thermolysis of cellulose with α-amino and α-hydroxy acids produces several aromas that potentially can be interesting for food and cosmetic industries. Starch and pectin also generate aromas on heating with α-amino acids.

Depolymerization of starch gives dextrins — one of the most useful products in the food industry. It is carried out either on the proton catalysis and/or thermally. Chemical dextrinization proceeds in the presence of hydrochloric acid, which is a superior catalyst, but also other acids, including carboxylic acids dextrinize starch. Depending on the reaction conditions, products of varying degrees of depolymerization are available. The simplest, behavioral depolymerization is achieved by heating aqueous starch suspension above its gelatinization temperature, followed by drying of the suspension. So-called pregelatinized starch becomes water-soluble due to destruction of in-matrix hydrogen bonds that hydrophobize starch. The depolymerization can lead up to D-glucose to give starch syrups. Contrary to chemical dextrinization producing white and yellow, canary, or thin- and thick-boiled dextrins, thermal dextrinization gives so-called British gums. The thermal rupture of glycosidic bonds is assisted by reversion leading to novel $1 \rightarrow 6$ bond formation. The acid-catalyzed dextrinization, only to a small extent, evokes contamination of products with oligosaccharide decomposition products. However, in the thermal dextrinization, the formation of a variety of contaminants cannot be avoided. Other physical energy sources also lead to dextrinization. High pressure up to 10^6 Pa results in destruction of starch macrostructure. Some depolymerization symptoms and assisting reversion could be observed when pressure above 1×10^6 Pa was applied.

Ionizing radiation depolymerizes starch, and such harmful by-products are formed as formaldehyde, acetaldehyde, glyoxal, formic acid, acetic acid, and five- and six-membered heterocyclic compounds. The up to 10 kGy dose is safe. The ultraviolet radiation, usually in the presence of sensibilizing metal oxides, causes depolymerization even up to water and carbon dioxide.

Chemical dextrinization proceeds according to an ionic mechanism, while thermolysis produces very stable radicals in high concentration. The first dextrinization step, dehydration, takes place at 100–160°C. The extent of random hydrolytic scission of the glycosidic bonds depends on the water content in starch. This step is common for both chemical and thermal processes. Further changes begin around 260°C. They are different for both kinds of procedure.

Chemical Modifications of Polysaccharides without Attempted Depolymerization

Polysaccharides offer practically the same kind of reactivity as monosaccharides. The reactions that occur in monosaccharides on the anomeric carbon atoms are limited to the first, terminal saccharide unit. Such minute modifications of polysaccharide chains evoke a clear response in some of their properties. For instance, rheological properties of starch gels vary after modifications carried out on one of its 1,250 glucose units.

The chemical modifications of starch also meet limitations. Steric hindrances of the reaction sites, as well as solubility and penetration of reagents, affect the degree of transformation.

Thus far, polysaccharides chemically modified for the food industry are starch, cellulose, and pectin. If biodegradable plastics are taken into account, the use of hemicelluloses may potentially be considered.

Chemical modification of starch involves oxidation, esterification, etherification, and complex formation. Metal derivatives might have some significance as carriers of bioelements for nutrition and therapy. The oxidation of starch to starch dialdehyde with periodates has little, if any, significance in the food industry, contrary to starch oxidized to carboxy starch, i.e., aldaric functions in the terminal and gluconic acid moieties in the chain. Practically important modifications should leave no more than one carboxylic group per each twenty-fifth glucose unit. The oxidation should be carried out with a possibly high depolymerization degree and the formation of the smallest possible number of the terminal anomeric carbon atoms. Such oxidation is provided by sodium hypochlorite. The 6-hydroxymethyl groups only randomly are oxidized to carboxylic groups. Such oxidation gives low viscosity sols and gels of improved transparency. Nitrogen oxides mainly turn the 6-hydroxymethyl group into a carboxylic group.

Starch can be esterified with organic and inorganic acids. These reactions are carried out in alkaline media using such reagents as acid anhydrides, including cyclic dioic acid anhydrides. The reaction mixtures are either stirred at room temperature, heated, or extruded. Phosphorylation can be carried out with sodium hydrogen phosphate. Only these preparations are utilized in foods that have a low, 0.2 to 0.0001 degree of esterification, although derivatives with a higher degree of substitution (DS) are readily available and utilized for nonnutritive purposes. Another type of starch acetylation deals with its transesterification with vinylacetate.

The most important and widely used starch ethers are those prepared using monochloroacetic acid (carboxymethyl starch), ethylene oxide (hydroxymethyl starch), and propylene oxide (hydroxypropyl starch). The syntheses proceed in alkaline solutions. If the dissociation leaves a negative charge on starch, various anionic starches, e.g., sulfonated and phosphorylated starches, carboxymethyl starch, and starch acylated with dioic acid cyclic anhydrides, are obtained. If the dissociation produces the positive charge on starches, their nature is cationic; e.g., tetralkylammonium starches, protonated amino starches are prepared. The degree of substitution in these modifications for their utilization in food processing should be as low as in starch esters. Although all three hydroxyl groups of glucose units can be esterified and etherified, that at C6 seems to have a priority, and the access of reagent is, perhaps, an important factor of that selectivity.

Starch is a very important complexing agent for inorganic, as well as organic, gases, liquids, and solids. Three types of complexation can be observed: (1) inclusion complexes inside of the amylose helix and eventually short helices of amylopectin as well as in sites of amylopectin branches, (2) surface sorption complexes, and (3) capillary complexes. All of them exist in a natural, native form; they can also be formed on several common operations of food processing, i.e., dough formation, beating of foam, scrambling yolk with sugar, etc. Formation of inclusion complexes of starch, actually amylose complexes, becomes a more common method of protecting some volatile, as well as oxidizable, food components. Such complexes are significant for food texturization and stabilization. Cationic starches are used as detergents and cellulose pulp components, whereas anionic starches readily form complexes with proteins. Such complexes are attempted to be utilized as biodegradable plastics and meat substitutes.

Modifications of cellulose for the use in food industry are limited to esters, mainly cellulose acetate for special membranes for treating water and fruit juices, and ethers. Methyl- and methylhydroxypropyl-celluloses are available in the Williamson reaction of cellulose with methyl iodide in alkali and/or subsequent etherification with propylene oxide also in alkali. Carboxymethylcellulose is the product of the reaction of cellulose with chloroacetic acid. Products of a degree of substitution from 0.3 to 0.9 are available. All 2, 3, and 6 hydroxy groups do react. Modified and derivatized celluloses have found their application in the food industry as nondigestible, ballast components of low-calorie meals.

Cross-Linked Starches

Starch cross-linking frequently occurs when it is esterified or etherified with bi- and polyfunctional reagents, e.g., $POCl_3$, polyphosphates, or dianhydrides of tetraionic acids. Such starches are mainly texturizing agents for their enhanced water binding, lower aqueous solubility, higher pH, and shear force stability.

Retrogradation is a very common reaction leading to enhanced molecular-weight systems. It is a property of amylose, but not amylopectin gels, that irreversibly recrystallizes into insoluble precipitate. Retrogradation is manifested by dendrite formation in the gel, as well as by bread staling. Low temperature around the freezing point, high amylose concentration, and polar gel additives favor retrogradation. The phenomenon is due to decoiling of the amylose helix into straight chains that undergo reorientation via hydrogen bond formation and form micelles. It is accompanied by conformational changes. The V-type amylose aggregates turn on retrogradation into B-form structures. The retrogradation affinity depends on starch variety, i.e., on the amylose/amylopectin ratio and, to a certain extent, residual native starch complexes present in starch. The affinity decreases in the order of potato > corn > wheat starch. Waxy maise starch does not retrograde at all because it contains approximately 98% of amylopectin. Retrograded starch is utilized as a component of low-calorie food.

ENZYMATIC TRANSFORMATIONS OF CARBOHYDRATES

With few exceptions, enzymatic processes in carbohydrates are degradative. Enzymes are used in form of pure or semi-pure preparations or their producers, i.e., corresponding microorganisms. Currently, semi-synthetic enzymes are also in use. The alcoholic fermentation is the most common way of utilization of monosaccharides and sucrose. The lactic acid fermentation is another important enzymatic process. Lactic acid bacteria metabolize mono- and disaccharides into lactic acid. This acid has a chiral center, thus either D($-$), L($+$) or racemic products can be formed. In the human organism, only L($+$) enantiomers are metabolized, whereas D($-$) enantiomers are concentrated in blood and excreted with urine. Among lactic acid bacteria only *Streptococcus* shows specificity in the formation of particular enantiomers and only L($+$) enantiomers are produced.

Enzymatic reduction of glucose-6-phosphate into inositol-1-phosphate with cyclase and reduced NAD coenzyme followed by hydrolysis with phosphatase presents another nondegradating enzymatic process on hexoses.

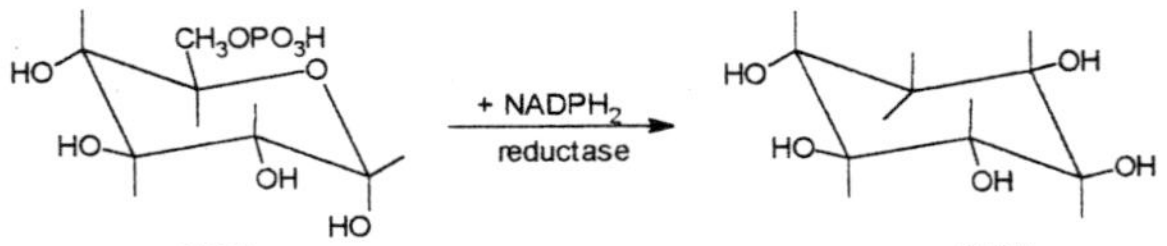

(4.80)   + NADPH$_2$ reductase   (4.81)

Inositol plays a role of a growth factor of microbes. Its hexaphosphate, phytin, resides in the aleurone layer of wheat grains.

There are also known bacteria that polymerize mono- and oligosaccharides. *Luconostoc mesenteroides* polymerizes sucrose into dextran — an almost linear polymer of about 400 α-D-glucose units. Dextran is also generated in frozen sugar beets. This creates difficulties in sugar manufacturing if the beets have to be stored in Europe at low temperature in December. Dextran serves as a blood substitute and chromatographic gel (Sephadex). Other polysaccharides synthesized by bacteria are levan, pullulan, a polymer of α-D-glucose, and xanthan.

The enzymatic oxidation of sucrose with glucose oxidase to D-glucose-δ-lactone consumes oxygen dissolved in beer and juices. This process decreases the rate of undesirable color and taste changes in these products; which are due to oxidation.

All essential enzymatic polysaccharide transformations deal with degradation. In the case of cellulose, this degradation leads to glucose, whereas starch, amylose, and amylopectin are not necessarily so deeply

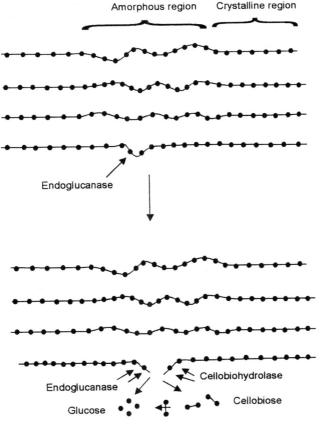

Figure 4.2 Enzymatic decomposition of cellulose.

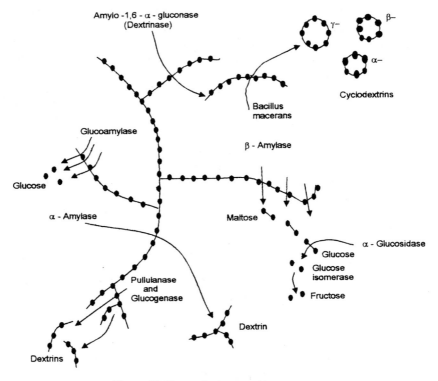

Figure 4.3 Enzymatic decomposition of starch.

degraded. Synthesis of cyclodextrins (cycloglucans, Schardinger dextrins) presents a special case.

Contrary to acid-catalyzed degradation, the enzymatic degradation of cellulose attacks fibrils located in the borderline between their amorphous and crystalline regions.

Complexes of cellulolytic enzymes composed of endoglucanase, exogluconase (glucohydrolase), celobiohydrolase, and celobiase (β-glucosidase) have to be used for the cellulose fermentation. They enter the fermentation in its different stages (Figure 4.2).

There are several amylolytic enzymes capable of starch degradation. They provide high specificity (Figure 4.3).

SACCHARIDES IN CEREALS

Starch is isolated from grains for its specific properties, e.g., granule size and susceptibility to physical and chemical modifications. The susceptibility of starch varieties to modifications results from the structure of starch granule and composition of grains. Proteins (7–13%), lipids (1.5–6%), other carbohydrates (5–23%), and minerals (1–3%) reside in the granules as native complexes that, only randomly, are separable to get pure starch. The most common sources of starch in various regions of the world are potato, maize, cassava (manioc, tapioca), and rice.

The starch reactivity with ammonia and proteins suggests that the carbohydrate-to-protein ratio, as well as carbohydrate type, may influence the suitability of sources for a given type of processing. The composition of the protein component is less important. These circumstances open a route to improvements of lower grade sources such as tobacco of too high a protein content, or cocoa and coffee seeds of either too low or too high a carbohydrate content. The thermal processing of wheat and rye flour under ammonia adds a mushroom-like flavor and aroma to such flours.

Cellulose, particularly, hemicelluloses, pectin, and mucus present in cereals are valuable components of food as so-called food fiber necessary for a proper peristaltic intestine action.

FUNCTIONAL PROPERTIES OF CARBOHYDRATES

TASTE

Except for sucrose, the sweetness decreases with the number of monosaccharide units in the oligosaccharide (Table 4.3) because only one monosaccharide unit interacts with the mucoprotein of the tongue receptor.

A derivatization of carbohydrates can specifically increase the sweetness. Because several nonsaccharide, natural and synthetic, powerful sweeteners are now available, such sweet derivatives have no practical significance.

The following carbohydrate sweeteners are in common use:

- D-glucose: Because of fast resorption, it is an energy source for children and patients as injection and infusion fluids. It is a sweetener for beer, beverages, and chocolate. Its digestion requires insulin. It causes tooth cavities.
- D-fructose: It is the most readily water-soluble sugar. It does not crystallize from stored juices. Because of hygroscopicity, it retains moisture in sugar-preserved food, for instance, in candized fruits, and intensifies their flavor and aroma. Its aqueous solutions have lower viscosity as sucrose solutions, but it increases the viscosity of beverages. Such solutions have lower energetic value than corresponding solutions of sucrose. It is the sweetener for fruits,

TABLE 4.3. Relative Sweetness (RS) of Various Substances in 10% Aqueous Solutions (sucrose = 1.0).

Substance	RS
Sucrose	1.00
β-D-Fructopyranose	1.80
Inverted sugar	1.30
α-D-Glucopyranose	0.70
β-D-Glucopyranose	0.80
α-D-Mannopyranose	0.30
β-D-Mannopyranose	bitter
D-Galactopyranose	0.32
Maltose	0.32
α-D-Lactose	0.20
β-D-Lactose	0.30
D-Galactosucrose	Sweetless
Raffinose	0.01
Stachyose	0.10
1'-Chloro-1-'desoxysucrose	0.20
4-Chloro-4-desoxysucrose	0.05
6-Chloro-6-desoxysucrose	Bitter
6'-Chloro-6'desoxysucrose	0.20
1',4-Dichloro-1',4-didesoxygalactosucrose	6.00
1',6'-Dichloro-1',6,-didesoxysucrose	5.00
6,6'-Dichloro-6,6'didesoxysucrose	Sweetless
1',4,6'-Trichloro-1',4,6'-tridesoxygalactosucrose	20.00
1',4,6,6'-Tetrachloro-1',4,6,6'-tetradesoxygalactosucrose	10.00
1',4,5,5'-Tetrachloro-1',4,6,6'-tetradesoxysucrose	2.00
Mannitol	0.40
Sorbitol	0.60
Xylitol	0.85–1.2
Honey	0.97
Molasses	0.74
Saccharin	200–700
Cyclamates	30–140
Aspartame	200
Neohesperidin dihydrochalcone	2,000

candies, ice cream, beverages, yogurt, juices, puddings, and tonizing drinks. It neither evokes nor accelerates tooth cavities. It accelerates ethanol metabolism. In the organism, D-fructose metabolizes into glycogen.

- lactose: It is present in mammalian milk (4.8–5.1%). It is sparingly water-soluble (20% at room temperature) and is utilized as a carrier of other sweeteners. It improves the flavor, produces a good image of food processed in microwave ovens, and improves the taste of dairy products.

- sucrose: It is the most common sweetener for its pleasant taste. Sucrose is widely used as a preservative for marmalades, syrups, and confitures. Its 30% solutions do not ferment, and 60% solutions are resistant to all bacteria but *Zygosaccharomyces*.
- maltose: Its sweetness is mild and pure. Its solutions have low viscosity. Maltose is slightly hygroscopic, and its color is stable regardless of temperature.
- starch syrups: They are the products of starch saccharification. Depending on the degree of starch conversion, various products are available. The first sweet product from the saccharification, maltotetraose syrup, has the sweetness of one-fifth that of sucrose. The syrups have high viscosity; they are water-soluble and slightly hygroscopic, do not retrograde, and are readily digested. Extended saccharification produces maltose syrups. Their viscosity is lower and thermal stability is good. Further saccharification leads to glucose syrup; which may be isomerized to give fructose syrups (Table 4.4). Glucose syrups are sometimes hydrogenated to give a product of RS = 0.2. This process is introduced to obtain the product of higher chemical and enzymatic stability, reduced tendency to crystallization, and better solubility. D-Sorbitol and mainly D-glucose can form polycondensate on citric acid catalysts. Apart from its sweetness and low energetic value (17.5 kJ/g), texturizing and filling properties are practically utilized. The syrups are designed for the manufacture of soft candies, pralines, and chewing gums.
- malt extract: It is an aqueous extract of barley malt. Apart from proteins, enzymes, and mineral salts, it contains 4–5% of sucrose and, in addition, some fructose, glucose, and maltose. It is utilized in candy manufacture and as the nutrient for baking yeast.
- maple syrup and maple sugar: They are manufactured from juice

TABLE 4.4. **Saccharide Content (%) in Various Starch Syrups.**

	Syrups				
	Glucose Conversion				
Saccharide	Low	High	Very High	Maltose	Fructose
Glucose	15	43	92	10	7–52
Fructose					42–90
Maltose	11	20	4	40	4
Higher saccharides	48	13	2	26	3–6

of *Acer saccharium* maple tree. The syrup contains 98% of saccharides, being sucrose in 80–98%.

- sugar alcohols: They are perfectly water-soluble, soluble in alcohol, and more stable at low and high pH than saccharides. Their sweet taste lasts for a prolonged time and is accompanied by cool impression. Their metabolism does not require insulin. Their energetic value reaches 17, 17, and 8.5 kJ/g for D-sorbitol, D-xylitol, and D-mannitol, respectively.
- compounded saccharide sweeteners: They are based on sucrose. Their sweetness is improved by blending sucrose with fructose, as well as fructose and glucose (1:1 mixture). Sometimes, sucrose is also blended with sugar alcohols. The latter is also combined with starch syrups and sometimes malic acid. Low energetic sweetener is available from blending starch syrup with sodium phosphate and maltol.
- honey: It is a natural product. Its composition depends on the harvest time, geographical origin, and kind of flowers from which the nectar was collected. Fructose, glucose, and maltose constitute approximately 90% of the total sugar content. There is also a rich variety of free amino acids and other organic acids, minerals, pigments, waxes, pollen, and enzymes. Honey also contains about 120 aroma substances. It may contain toxic components if the nectar is collected from poisonous plants, although there are several poisonous plants that give nonpoisonous honey. Honey was for millennia utilized as a sweetener in either plain form or as an additive to alcohols, water, tea, and milk. In some countries, it is fermented into honey-flavored wine (mead). Honey may, however, have some allergizing properties.

COLORANTS

Caramel is a brown colorant and, to a certain extent, also a flavoring agent prepared by burning sugar (caramelization). As a food additive, its manufacture, types, and properties are standardized. When the caramelization proceeds without any catalyst at 200–240°C, plain caramels are obtained, the color intensity of which is relatively poor. These caramels serve as flavoring additives, rather than colorants. Caramels from catalyzed processes requiring lower temperature (130–200°C) have higher color intensity. They are typical brown food colorants. The standardized catalysts are caustic soda, caustic sulfite, ammonia, and ammonia combined with sulfite. As far as the color intensity is considered, ammonia is the superior catalyst. Unfortunately, there is a concern about health properties of ammonia caramels. Neurotoxicity of 4(5)-methylimidazole, which is formed on such caramelization, puts a limit on the daily uptake of that caramel. In laboratory tests, some α-amino acids and their sodium and magnesium salts were found to be effective caramelization catalysts, additionally providing a specific flavor of caramels. Another concern deals with mutagenicity of caramels that contain radicals. A recent study has shown that caramels are nonmutagenic.

TEXTURIZATION

Saccharides, particularly polysaccharides, form hydrocolloids, which build up their own macrostructure. They give an impression of jelly formation, thickening, smoothness, stabilization against temperature and mechanical shocks, aging, and resistance to sterilization and pasteurization.

Carbohydrates are capable of complexation with a variety of compounds. It implies that the texturizing role of saccharides depends on their concentration, reaction conditions (temperature, pH, reaction mixture composition), lipid content, and protein structure. Because saccharides and polysaccharides form complexes with one another, such carbohydrate mixtures can also show increased viscosity and adhesiveness. Therefore, they can be employed in controlling the texture of foodstuffs and designing novel, fully digestible food glues. The time necessary for the development of the effect and its stability are also essential. The stability of such gels is satisfactory in the pH range of 4.0–7.0. Cross-linked starch is frequently used as the texturizing agent. The degree of cross-linking is an important factor. The viscosity of gels is not proportional to the cross-linking degree. Among cross-linked starch modifications, those esterified with phosphoric acid are particularly favored. The water-binding capacity of mono- and diesters is affected by cross-linking and the degree of substitution (Table 4.5). Phosphorus content in starch phosphates is limited by food laws of particular countries and does not exceed 0.5% in starch monoester and 0.14% in starch diester.

Starch sulfate ester is used as a thickener and emulsion stabilizer. It is a typical anionic starch used as a component of anionic starch–protein complexes constituting meat substitutes. Other anionic starches, as well as pectins, alginic acid, and carrageenan, may also be utilized. The protein suitable for texturization should have a molecular weight of 10–50 kDa. Proteins of lower molecular weight are poor fiber builders, while proteins of larger molecules are too viscous and can gelate above pH 7. The protein-to-anionic starch ratio depends on the texturization method. Among many potentially available modified starches, only a few of them are accepted by the food laws. Some restrictions are put on the method of the manufacture and purity of such products.

TABLE 4.5. Water-Binding Capacity (WBC) (g/g of dry substance) and Aqueous Solubility (AS) (%) of Starch of Various Origin as Well as Mono- and Di-Starch Phosphates at 60°C.

Starch Variety	Free		Monoester		Diester	
	WBC	AS	WBC	AS	WBC	AS
Potato			(0.038)		(0.0011)	
	13	7	87	70	17	6
Maize			(0.040)			
	2	1	54	66	2	1
Wheat			(0.025)			
	7	4	52	49	6	4
Rye			(0.035)			
	8	4	43	72	9	6
Triticale			(0.023)			
	8	2	51	61	7	4
Waxy maize			(0.040)			
	2	2	35	90		

*The degree of substitution with phosphorus is given in parentheses.

Because of carboxylic groups in some polysaccharides, a maximum effect is achieved in the presence of sodium, potassium, and calcium cations. For this and other reasons, the properties of pectin are sensitive to the degree of methylation and esterification. Thus, pectins are grouped into highly methylated or low esterified (DE > 50%) and low methylated or low esterified (DE < 50%). The latter is available by hydrolysis of highly substituted products. When high viscosity is required, gum tragacanth is superior, whereas gum Arabic affects the viscosity to a lesser extent. In several cases, very good results come from the combination of saccharides.

ENCAPSULATION

α-, β-, and γ-Cyclodextrins are the most effective compounds for microencapsulation of food components volatile and sensitive to oxidation, light, and temperature. These cyclodextrins form toruses (Figure 4.4) of the following dimensions:

	ϕ_1	ϕ_2	h [A]
α	5.7	13.7	7.8
β	7.8	15.3	7.8
γ	9.5	16.9	7.8

Edges of the torus cavity are populated by secondary and primary hydroxyl groups at the upper and bottom torus edges, respectively. All hydroxyl groups are situated on the external surface and, therefore, cyclodextrins are hydrophilic, good water-soluble compounds with hydrophobic cavities. These facts determine the applicability of cyclodextrins. They form inclusion complexes with guests being nonpolar compounds and anions. The guest dimension is an additional factor, which governs the nature of inclusion complexes and selectivity in their formation. Commercially available cyclodextrins are, in fact, their inclusion complexes with water. Water is reversibly included and can readily be expulsed by other guests that are similarly reversible complexed. Comparative studies on the encapsulation effectiveness of glucose and β-cyclodextrin against several guest molecules show the superiority of the latter. However, D-glucose shows some effectiveness as the material for microencapsules. Encapsulation is also exhibited by gum Arabic and arabinogalactan from starch.

Known complex formation of starch has also found its application in similar areas as that for cyclodextrins. Contrary to cyclodextrins that employ torus cavities for the complexation, starch may form three types of complexes simultaneously: (1) surface, (2) capillary, and (3) inclusion complexes with an involvement of the amylose helix.

OTHER CARBOHYDRATE APPLICATIONS

Carbohydrates in food processing have found other applications. Thus, the complex of carbon dioxide in α-cyclodextrin is proposed as a baking powder. Marmalades, juices, confitures, jellies, etc., are effectively preserved by addition of sugar at relatively high concentration. Osmotic

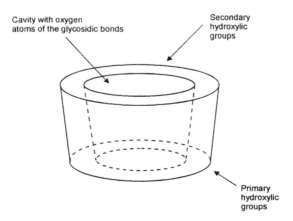

Figure 4.4 Scheme of cyclodextrins.

phenomena are involved. Due to competition of microorganisms and preserved foodstuffs for water molecules, the microorganism tissues undergo plasmolysis.

BIODEGRADABLE PLASTICS

There is a growing interest in biodegradable plastics as packing and wrapping materials, containers, plates, forks, knives, spoons, cups, and table blankets in fast-food restaurants, as well as in materials of very high water-binding capacity, up to 3,000 g $H_2O/1$ g of plastics. Apparently, the simplest biodegradable plastics could be prepared of starch solely by compression up to 10^9 Pa, provided starch was moisturized up to its natural water-binding capacity (20%). However, neither mechanical properties nor bioresistance of such material would satisfy the users.

Following the idea of full biodegradability of plastics, attention has been paid to the compositions of plain carbohydrate with unmodified protein, modified carbohydrate with unmodified protein, plain carbohydrate with modified protein, and modified carbohydrate with modified protein. In all cases, emulsifiers such as glycerol and fillers are added. Such compositions are processed in order to form carbohydrate–protein complexes. An electrical compatibility of the components should be reached in order to afford the best functional properties of the product. If the carbohydrate component is not naturally anionic, it should be made anionic. DS of such derivatizations should be low (< 0.1), and the derivatization should neither increase the hydrophilicity of the product nor, if possible, decrease its molecular weight.

Modifications of the protein are mainly random acylations. Components blended in varying proportions are extruded and subjected to either temperature or pressure molding. Electrochemical complex formation on anode is also described.

Several attempts have been made to modify polyethylene into the biodegradable material. Thus, it is usually blended with natural components such as starch, cellulose, wood, and/or proteins. Such modified polyethylene foils contain 6–15% of starch either in the solid or gelatinized form. Polyurethane foams as thermal insulators and packing materials contain up to 20% of starch. When starch is utilized in such plastics and in copolymers of ethylene with either vinyl chloride or styrene, the contribution from starch reaches 50%. The same amount of starch is met in its copolymers with acrylic acid, which are excellent water binders.

CHAPTER 5

Lipids[1]

JAN SAWICKI
BARTŁOMIEJ ŚLUSARSKI

PART 1: CHEMICAL AND FUNCTIONAL PROPERTIES

INTRODUCTION

L IPIDS are one of the important groups of organic compounds of a great physiological role and present in all food products. They are also a good source of energy (37.7 kJ/g). Lipids are composed of many classes of compounds. Most of them are apolar, and others have very polar character. In living organisms, lipids play a unique role in the structure of biomembranes and in transporting biologically active compounds, e.g., vitamins. The general classification of lipids is given below.

(1) Simple lipids—esters of fatty acids with glycerol
   - fats—esters of fatty acids with glycerol
   - waxes—esters of higher fatty acids with alcohols other than glycerol
(2) Compound lipids—esters of fatty acids with glycerol containing also other groups
   - phospholipids/phosphatides—lipids containing phosphoric acid residues as mono- and diesters
   - glycerophospholipids—derivatives of glycerophosphoric acid having at least one O-acyl, O-alkyl, or O-alkenyl group adjacent to the glycerol moiety
   - sphingolipids—derivatives of 1-phosphoceramide
   - glycolipids—containing a carbohydrate adjacent to lipid moiety

---

[1]Part 1 is authored by Jan Sawicki, and Part 2 is authored by Bartłomiej Ślusarski.

- glycoglycerolipids—glycolipids containing one or more glycerol moieties
- glycosphingolipids—compounds containing the carbohydrate sphingoide

(3) Derived lipids—derivatives of simple and compound lipids
- fatty acids
- alcohols
- hydrocarbons

Natural lipids are known as fats or oils. Fats are solid, while oils are liquid at ambient temperature. Lipids of animal origin are called fats or greases, and plant lipids are referred to as oils. Compound lipids and free fatty acids (FFA) are eliminated in the refining process of vegetable oils. Animal fats are designated for edible purposes, mostly in a crude state. Only a small part of plant oils is utilized in a crude state, e.g., virgin olive oil.

THE CHEMICAL CONSTITUTION OF MAJOR LIPID CLASSES

Fatty Acids (FA)

Nomenclature

FAs are the basic units of every lipid in food products. FAs are straight-chain aliphatic compounds with an even number of carbon atoms. Classification of FA is based on the length of carbon chain or the degree of unsaturation. The most common classification is saturated, mono-, di-, and polyunsaturated acids. The nomenclature of FA is a mixture of trivial and systematic names. The systematic name of a saturated FA is adopted from the parent hydrocarbon and the final "e" is replaced by "oic." When FAs are unsaturated, the parent hydrocarbon has a suffix "-ene," "-diene," or "-triene" with "-oic" for the acid group. Thus, octadecane and its saturated analogs give names for four FAs: octadecanoic, octadecaenoic, octadecadienoic, and octadecatrienoic. The position of the double bond and its geometrical configuration is determined according to the rules of the Geneva Convention. The carboxyl group position is designated at 1. The systematic names are too long in practical use, so the shorthand notation was adopted. The term C18:1(9c) is used for octadeca-9 enoic acids. The suffix "c" is omitted when the configuration of double bond is *cis*. The position of the double bonds in polyunsaturated acids is indicated with the suffix "n," which refers to the ending $-CH_3$ group and in that system is designated as 1. Oleic acid is oleic (n-9). Polyunsaturated FAs (PUFAs) belong to the n-9 and n-3 families.

TABLE 5.1. Saturated Fatty Acids.

Number of Carbon Atoms	Trivial Name	Systematic Name	Melting Point (°C)
4	Butyric	Butanoic	−5.3
6	Caproic	Hexanoic	−3.2
8	Caprylic	Octanoic	16.5
10	Capric	Decanoic	31.6
12	Lauric	Dodecanoic	44.8
14	Myristic	Tetradecanoic	54.4
16	Palmitic	Hexadecanoic	62.9
18	Stearic	Octadecanoic	70.1
20	Arachidic	Eicosanoic	76.1
22	Behenic	Docosanoic	80.8
24	Lignoceric	Tetracosanoic	84.2

Saturated Fatty Acids

Saturated FAs are, with a few exceptions, solid at ambient temperature. The lauric acid contents in coconut and palm kernel oils is 45–50% and 45–55%, respectively. Miristic acid in these oils is 15–19% and 15–18%. Palmitic acid is present in all vegetable oils. In palm and cottonseed oils, FAs are found in quantities of 35–45% and 22–28%, respectively. Milk fat and some fish oils contain 20–30% and 12–20% palmitic acid, respectively. Stearic acid is present in tallow 15–30% and in cocoa butter ~35%. Short-chain acids (C_4–C_{10}) are found in milk fat.

Unsaturated Fatty Acids

The unsaturated fatty acids (UFAs) are composed of two groups— monounsaturated FAs and polyunsaturated FAs (PUFAs). The double bonds in UFAs are isolated by a methylene group ($-CH=CH-CH_2-CH=CH-$) and the configuration is mainly *cis*. The *trans* isomers of UFAs are also found in natural lipids, e.g., in butter in limited quantity, while in significant quantity in hydrogenated fats.

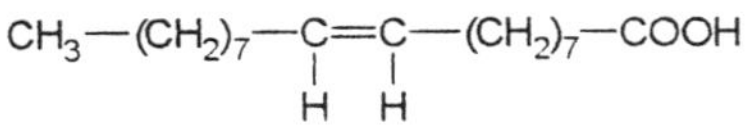

$$CH_3-(CH_2)_7-\overset{\displaystyle |}{\underset{\displaystyle H}{C}}=\overset{\displaystyle |}{\underset{\displaystyle H}{C}}-(CH_2)_7-COOH$$

Oleic acid 18:1 (9c)

(5.1)

$$CH_3-(CH_2)_7-\overset{\displaystyle H}{\underset{\displaystyle |}{\overset{\displaystyle |}{C}}}-\overset{\displaystyle H}{\underset{\displaystyle H}{\overset{\displaystyle |}{C}}}-(CH_2)_7-COOH$$

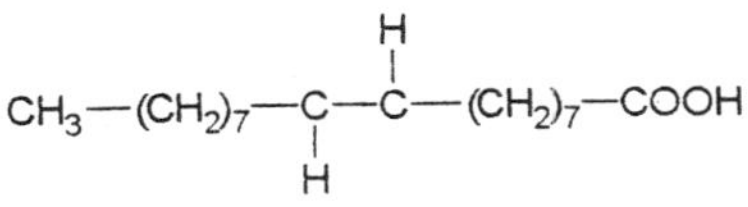

Elaidic acid 18:1 (9t)

(5.2)

TABLE 5.2. Monounsaturated Fatty Acids.

Number of Carbon Atoms	Trivial Name	Systematic Name	Melting Point (°C)
14	Myristoleic	cis-9-Tetradecenoic	−4.5
16	Palmitoleic	cis-9-Hexadecenoic	−0.5
18	Oleic	cis-9-Octadecenoic	16.3
20	Gadoleic	cis-Eicosenoic	23.5
22	Erucic	cis-13-Docosenoic	33.5
18	Elaidic	trans-9-Octadecenoic	46.5
22	Brasidinic	trans-13-Docosenoic	61.9

Oleic acid ($C_{18:1}$) is a dominant component in olive oil, in which its content is up to 75%. It is present in almost all fats. Its percentage in cocoa butter is ~40% and in lard and tallow up to 35–40%. Oleic acid at ambient temperature is liquid but at the same condition elaidic acid is solid. The melting point of *trans* erucic acid is 30°C higher than that of *cis* form. The *trans* isomers can replace solid FAs in fat mixtures. The physiological function of *trans* isomers is not finally established. PUFAs are present in greater quantity in vegetable and fish oils. Food products containing phosphatides are a good source of PUFAs. The conventional sources of PUFAs are limited. Recent developments in biotechnology make it feasible to produce PUFAs by microorganisms.

Acylglycerols

Acylglycerols consist of triacylglycerols (TAGs), diacylglycerols

TABLE 5.3. Polyunsaturated Fatty Acids.

Name of Acid	Structure	Family
Linoleic	18:2(9,12)	n-6
γ-Linolenic	18:3(6,9,12)	n-6
	20:3(8,11,14)	
Arachidonic	20:4(5,8,11,14)	n-6
	22:4(7,10,13,16)	
	22:5(4,7,10,13,16)	
α-Linolenic	18:3(9,12,15)	n-3
	18:4(6,9,12,15)	
	20:4(8,11,14,17)	
Eicosapentaenoic	20:5(5,8,11,14,17)	
	22:5(7,10,13,16,19)	
Docosahexaenoic	22:6(4,7,10,13,16,19)	

DAGs), and monoacylglycerols (MAGs). The fats of food products are composed mainly of TAGs. When TAGs are hydrolized, partial esters of glycerol and FFAs are formed. Fats of good quality that are used as food should have low contents of FFAs.

The determination of structure of TAGs is very difficult. When the three OH groups of glycerol are esterified with n different acids, the number of possible TAGs equals n^3. The determination of the stereospecific structure of TAGs is possible by the special use of hydrolytic selective enzymes. Thus, the differentiation between the 1 and 3 positions was possible (Brockerhoff, 1965). The position of FAs in triacylglycerols is given with the prefix "sn-." The conventional enzyme analysis does not distinguish between positions of FA at C-1 and C-3. Most of the results of TAG analysis in different oils were obtained by this method. According to Kartha (1951), the saturated and unsaturated FAs are distributed in a way to assure solubility of trisaturated TAG in fluid lipids. According to Gunstone (1964), there is a preferential acylation of the C-2 position by unsaturated C_{18}. Palmitic and stearic acid are located in the 1 and 3 positions. There are different types of FA distribution in TAGs of plant, animal, and microbial lipids. As the result of that specificity and composition of FA, the fats and oils have typical properties.

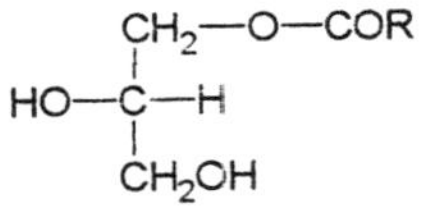

1 - acyl - *sn* - glycerol   1 - acyl - glycerol   3 - acyl - *sn* - glycerol

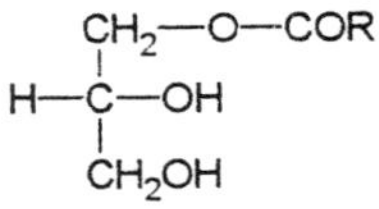

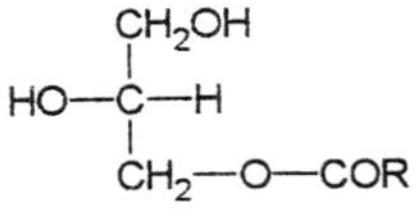

(5.3)   (5.4)   (5.5)

Waxes

Waxes are esters of FAs with higher aliphatic alcohols. Their contents in vegetable and animal fats is low. Only jojoba oil is a rich source of waxes. In food grade oils and fats, waxes are in trace quantity. Some fish oils are rich in waxes, especially the oil from orange roughy, a fish abundant in New Zealand fishing grounds. Deep skinning must be applied to remove the underskin layer of fat from the fillets. Waxes are separated from the oils by crystallization and subsequent filtration.

Compound Lipids

Phospholipids

Phospholipids or phosphatides, are very characteristic lipids for every

living organism. Their fatty acid composition is typical for the given
organism. The contents of phosphatides in raw vegetable oil is a few percent: in soybean oil, up to 3.5% and in canola oil, up to 0.5%. The contents of phosphatides in the lipids of fish is 0.4–0.8%, of egg yolk is
~20%, and of edible invertebrates is up to 30–40%. The phosphatides are
a source of phosphorus, unsaturated FAs, and aminoalcohols like choline.
Refined edible oils are free of phosphatides. Lard and tallow contain trace
amounts of phosphorus compounds.

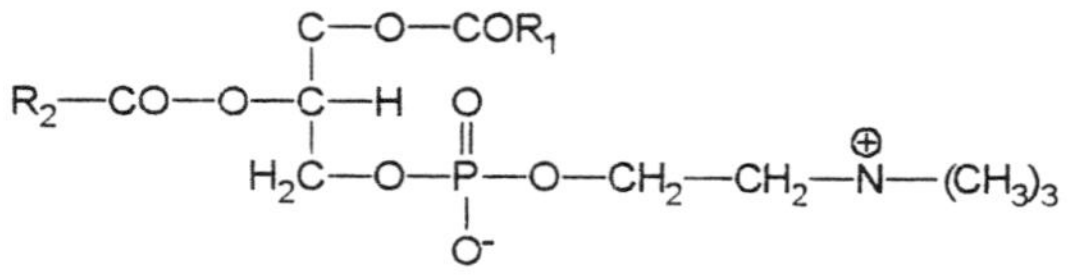

$$R_2-CO-O-\underset{\underset{H_2C-O-\overset{O}{\underset{\underset{O^-}{|}}{\overset{||}{P}}}-O-CH_2-CH_2-\overset{\oplus}{N}-(CH_3)_3}{\overset{|}{\underset{|}{C}}-H}}{\overset{C-O-COR_1}{\overset{|}{|}}}$$

Lecithin

(5.6)

Derived Lipids

Alcohols

Alcohols that are found in food lipids belong to sterols, aliphatic alcohols, tocopherols, and triterpenoic alcohols.

Sterols

Sterols found in food lipids belong to three groups: zoosterols, phytosterols, and mycosterols. The very characteristic representatives of these
groups are cholesterol, sitosterol, and ergosterol. The most important
dietary role is played by cholesterol. Cholesterol is found in animal food
products that have rather high contents of saturated FAs. Phytosterols are
in plant lipids, which are abundant in unsaturated FAs. The fish and
invertebrate lipids are rich in cholesterol and unsaturated FAs.

Sterols are crystalline compounds melting at about 150°C. In lipids,
they are in free or esterified forms. In animal fats, sterols are present in the
free form while in plant lipids, they are esters. Sterols can be separated
from the lipid mixture by saponification and subsequent extraction from
the soaps of so-called unsaponifiable matter. Sterols represent more than
half of unsaponifiables contents. During the refining of fats and oils, the
sterols are dehydrated, giving hydrocarbons. Detection of sterol hydrocarbons in fats and oils is a proof of refining.

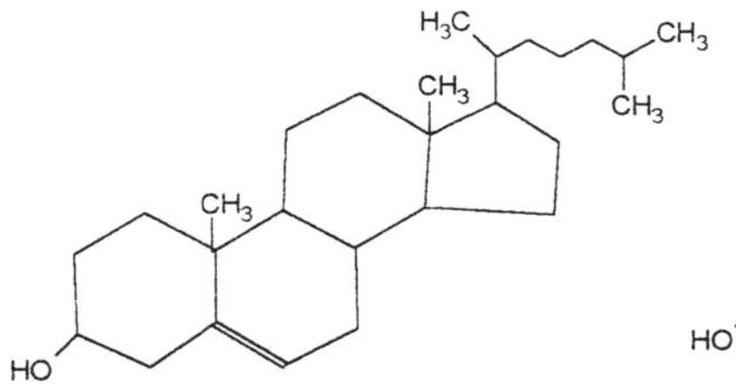

Cholesterol

(5.7)

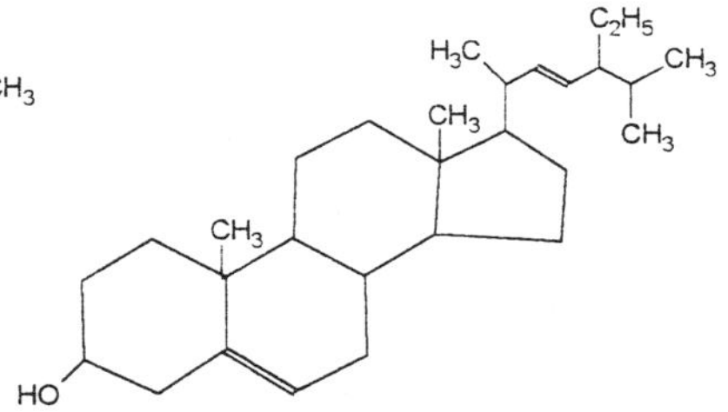

Stigmasterol

(5.8)

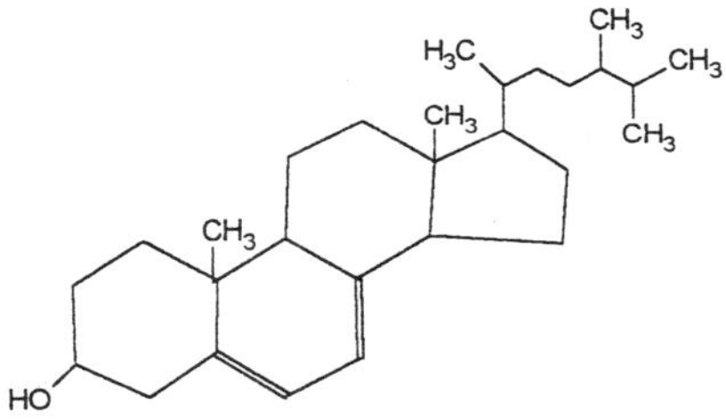

Ergosterol

(5.9)

Compounds of Vitamin E Activity

These compounds belong to isoprenoids. Their contents in plant oils is correlated with the unsaturation of FAs. They are composed of two groups: tocopherols and tocotrienols. They are natural antioxidants. The natural sources of tocopherols are plant oils and some by-products of oil refining. The tocopherols in oils are mixtures of different isomers. The highest contents of tocopherols are in soybean, rapeseed, and especially wheatgerm oils. During refining of oils, the tocopherols are partially eliminated, but still about 80–85 % of them remain. Thus, even refined oils are well protected against oxidation.

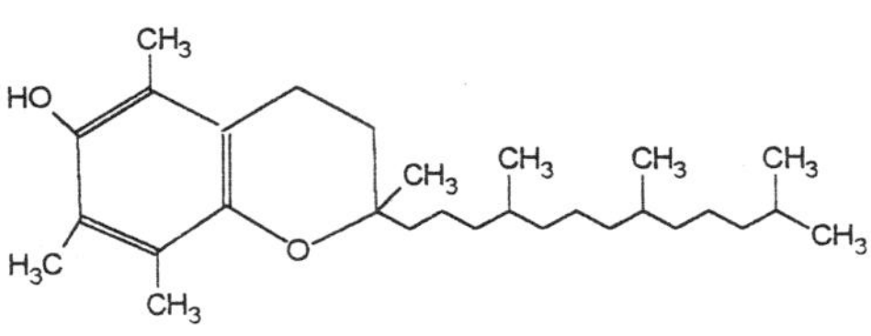

α - tocopherol

(5.10)

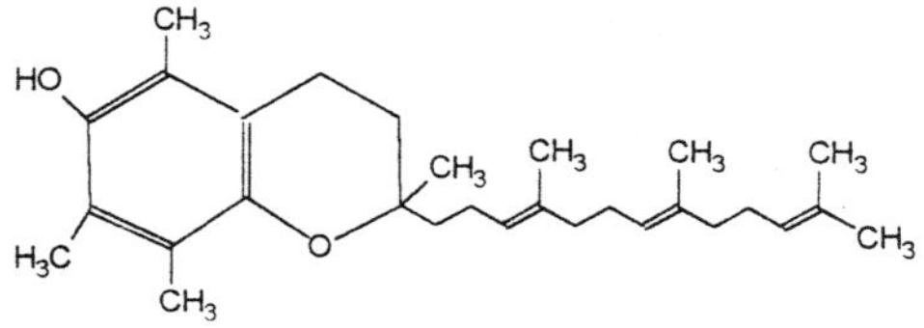

α_1 - tocotrienol

(5.11)

Triterpenoid Alcohols

Triterpenoid alcohols belong to methylsterols of tetra- and pentacyclic structure. In plant oils, there are triterpenoid alcohols and methyl sterols. The first have two methyl groups. The contents of triterpenoids in plant oils is negligible. These compounds are useful in identification of the purity of oils.

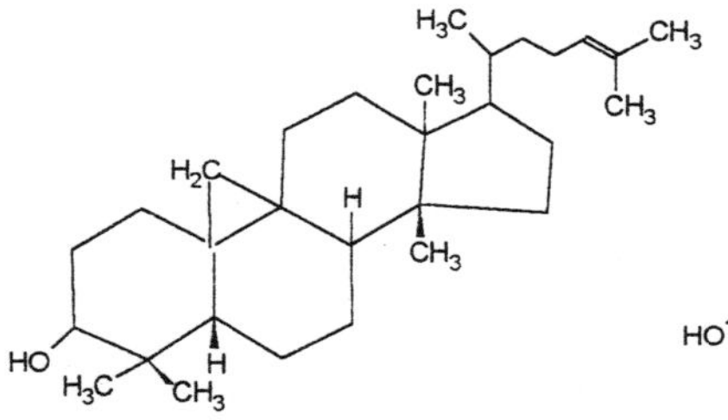

Cycloartenol

(5.12)

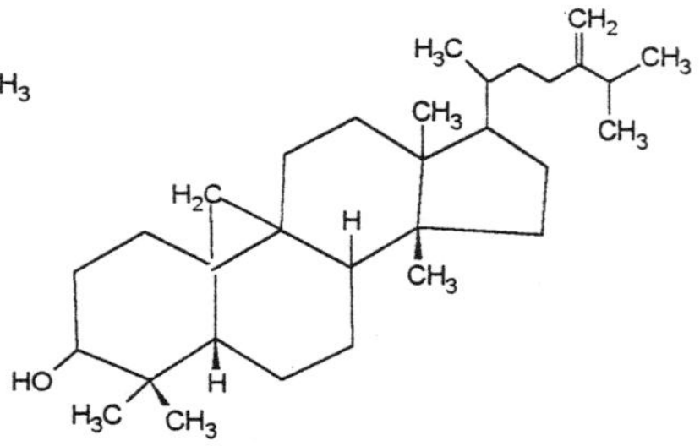

24 - methylene - cycloartanol

(5.13)

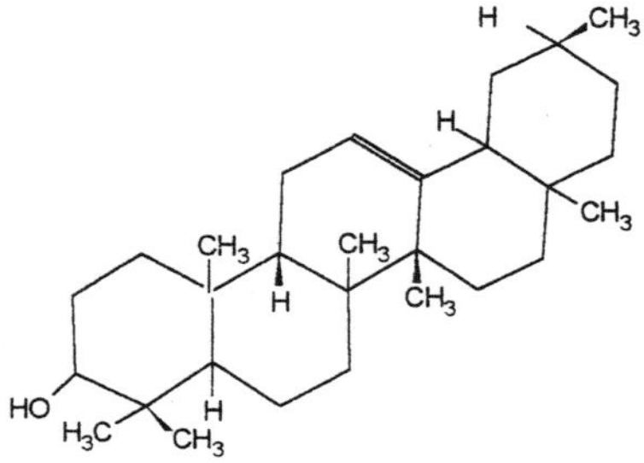

β - amyrin

(5.14)

Hydrocarbons

The other minor substances that are found in raw and refined oils are aliphatic alcohols, phytol, and hydrocarbons. The qualitative and quantitative analysis of minor substances is important for checking the purity of oils. Analysis of hydrocarbons is of value when contamination of oils with mineral oil is to be proved. The very characteristic, naturally occurring hydrocarbon of fish oils is squalene.

CHEMICAL AND ENZYMATIC MODIFICATION OF LIPIDS

Interesterification

The natural oils and fats have typical physical properties. Interesterification is used to modify these properties and to increase the functional value of fats as components of different fatty food products.

The acyl groups in TAG may exchange positions within the molecule or between molecules. Interesterification is catalyzed by NaOH, Na-K alloy, and CH_3ONa. The reaction is usually performed at a temperature of about 100°C. Traces of water should be eliminated from the reaction mixture. In proper conditions, the exchange reaction lasts only a few minutes. Interesterification can be accomplished in any fat or mixture of fats. The mechanism of reaction is the same in any case. Interexchange of acyls within one molecule occurs more rapidly than those between two molecules. The distribution of acids in TAG after interesterification is random or statistical. As a result of reaction, the chemical and physical properties are changed. The modified fats have modified properties such as plasticity and are valuable components in margarines, shortenings, and enrobing fats.

Directed interesterification is performed at low temperature, e.g., 40°C. In these conditions, some saturated TAGs are eliminated from the reaction mixture. This type of interesterification is used only to obtain very special fats like cocoa butter.

The new interesterification method is based on application of enzyme technology. Enzymes isolated from *Mucor miehei* can catalyze exchange of acyls in *sn*-1,3 position. The enzyme technology can be used for obtaining partial acylglycerols with higher yield than in the conventional method.

By applying the interesterification to some cheap fats like lard and tallow, they can be converted to valuable products. Natural lard and tallow may be used only in a limited number of food products. As the interesterified fats, they are valuable components of many fat products.

Hydrogenation

Hydrogenation of liquid fats and oils is performed to obtain plastic fats stable to oxidation. These products are used as components of margarine and shortenings. Partially hydrogenated oils are still liquid and can be used as salad oils. Because the most unsaturated FAs are transformed to saturated ones, partially hydrogenated oils are more resistant to heat and oxidation.

The hydrogenation is performed on an industrial scale at temperatures of 120–220°C. The pressure of hydrogen is 1–6 atm. Ni-catalyst is added in the quantity ~0.05%. The suspension of oil and catalyst is intensively mixed. Saturation of double bonds of acyls takes place on the surface of the catalysts. The catalysts of hydrogenation are metals as Ni, Cu, Cu−Cr, and Pt. The most critical parameters of reaction are the rate of the diffusion of hydrogen in the liquid phase and its adsorption, desorption, and other phenomena on the surface of the catalyst. In industrial practice, only edible oils are partially hydrogenated. During hydrogenation, saturation and isomerization of double bonds takes place. The selectivity of hydrogenation is determined as the rate of saturation of different acids: trienoic, dienoic, enoic. Isomerization is determined as the rate of formation of *trans* isomers. The formation of *trans* and positional isomers depends on the reaction of active centers of catalysts (Drozdowski, 1981). The saturation of double bonds proceeds by a state of half hydrogenated acyl. If further addition of a hydrogen atom is performed, the saturated acyl is obtained. Another possibility is the desorption of a half hydrogenated acyl with the elimination of a hydrogen atom. In this case, the regenerated unsaturated acyl is *trans* or *cis* isomer. Positional isomers are also formed by the same mechanism. The hydrogenation is selective when trienoic acids are converted to dienoic ones without changes of dienoic and enoic acids to more saturated ones. The contents of *trans* isomers and the degree of saturation depends in industrial processes on the properties of the catalyst and on the parameters of reaction. In drastic conditions of hydrogenation, undesirable reactions take place, e.g., hydrolysis, pyrolysis, and polymerization. The hydrogenated fats must be fully refined to eliminate traces of catalysts and of undesirable substances.

CHANGES OF FATS DUE TO STORAGE AND PROCESSING

Changes of Lipids in Food Raw Materials

Dry soybeans and rapeseeds are stored before processing. Conditions of storage and handling must be chosen to avoid some deterioration of lipids, proteins, and carbohydrates of the seeds. The complex deterioration of the seeds may cause the processing to be more difficult. The kind of lipid changes during storage are hydrolysis, autoxidation, and degradation of the pigments. Oils obtained from deteriorated seeds have high contents of FFAs and dark color. The palm oil is obtained from wet material. To stop the activity of enzymes, the fresh palm bunches are thermally treated.

Animal Origin Food Raw Materials

Animal fats are obtained from fresh raw materials, just after slaughter of the animal. In the thermal process of obtaining lard and tallow, the enzymes are deactivated and the fat is liberated. The very parameter of good quality of animal fats is the freshness of the raw materials.

Changes of Lipids during Processing

The Refining of Fats and Oils

The refining of fats and oils is carried out in many unit operations. The main purpose of refining is the elimination of undesirable compounds from the oil. During refining also, valuable compounds are completely or partly eliminated from the oil. The great problem in refining is the formation of artifacts. The side effects are limited and refining is the only operation that secures the fats of GRAS type. The main refining operations are given below.

Hydration

In the hydration step, the phosphatides are transformed into oil-insoluble substances. The hydrated phosphatides are easily separated from the oil by centrifugation. That operation is, to some degree, natural because during storage of crude oils, the formation of sediments composed mostly of phosphatides can be observed. The phosphatides isolated from the oil during hydration are dried and sold as lecithin.

Neutralization

The FFAs are neutralized by alkali. The concentrated water solution of NaOH is mixed intensively with the oil at a temperature of ~90°C. The solution of soaps is separated from the oil by centrifugation. In the drastic conditions of neutralizing, many substances are eliminated from the oils: FFAs, oxidized derivatives of lipids, traces of metals. The neutralized oil is lighter than the crude one.

Bleaching

Another term for this process is decolorizing or adsorption. The last term is the most adequate because, in this operation, the adsorption phenomena are responsible for the bleaching effect. Industrial bleaching consists of mixing the oil with ~1% of the adsorbent, e.g., bleached earth. The temperature of the operation is ~95°C. The adsorbent is separated from the oil by filtration. The bleached oil is free of polar substances and has lighter color than neutralized one. Artifacts of the bleaching are diene- and triene- of FAs in negligible quantity. There is a possibility to avoid formation of diene- and triene- of FAs by using neutral adsorbents. Bleaching is the main operation in the physical refining when FFAs are eliminated by distillation. The adsorption of undesirable substances is the basic operation when oil or fat should be only gently refined. Some gourmet oils are gently treated with adsorbents.

Deodorization

In this process, at about 200°C, the odor substances are distilled off at very low pressure. To facilitate the distillation of odorous substances, the process is performed with preheated steam. The same industrial installation may be used for deodorizing of neutralized oils as for the deacidification of the adsorbent treated oils. At high temperature of the process some side reactions take place, e.g., polymerization and interesterification. The parameters of industrial process are so established as to minimize the formation of artifacts. During deodorizing some biologically active substances, e.g., tocopherols, are partially eliminated from the oil.

REFERENCES

Brockerhoff, H. 1965. "A stereospecific analysis of triglycerides," *J. Lip. Res.* 6, 10–15.

Drozdowski, B. 1981. "The mechanism of heterogeneous catalytic hydrogenation of fatty oils," *Zeszyty Problemowe Postępów Nauk Rolniczych.* 211:47–59.

Kartha, A. R. S. 1951. *Studies of Natural Fats Vol. 1.* Published by the author. Ernakulam, India.

Gunstone, F. D. 1964. "The long spacing of polymorphic cristalline form of long-chain compounds with special reference to triglycerides," *Chem. and Ind.* 84–92.

PART 2: THE EFFECT OF REFINING AND HARDENING ON THE COMPOSITION OF FATS

INTRODUCTION

Fundamental differences occur between the chemical composition of virgin, refined, hydrogenated, and transesterified fat. The virgin oil is exactly the same as that present in the plant seeds. It is yielded using mechanical methods only (Codex Alimentarius Commission, 1983). The refined oil is obtained by extracting plant seeds with organic solvents, degumming, deacidifying, bleaching, and deodorization. These oils are applied for salads, sauces, and occasionally for short frying.

The plant oils are relatively inexpensive, but they are liquid. The consumer market also demands solid fats as butter replacers, for frying and baking, with long storage stability and adequate rheological properties. Such fats are produced from oils by transesterification or hydrogenation. The hydrogenation replaces the polyunsaturated fatty acids (FAs) with saturated and monounsaturated FAs. The transesterification reaction increases the melting temperature of the oils. During this process, the oils are totally hydrogenated until all double bonds are saturated, and in the second stage, the fat is transesterified with liquid refined oil, giving the final product. Such fats are used as an addition in margarine formulations and for frying purposes.

Apart from the intended changes, there are some other modifications of the fats, which take place during hydrogenation and transesterification and lead to the formation of *trans* FA isomers. The effect of these isomers on human health is under investigation. Thus, it is very important to know the fatty acid composition of fats used for human nutrition.

Examination of FA and TAG composition are the fundamental tasks in fats analysis.

METHODS OF DETERMINATION OF FATTY ACIDS COMPOSITION

FAs are analyzed most often as methyl esters using gas chromatography (GC). The esters are more volatile and less polar than the corresponding FAs. This improves the chromatographic separation. GC is applicable to separate and determine esters having four to twenty-four carbon atoms in the molecule according to their degree of unsaturation and carbon chain length. The complete analysis consists of several steps: esterification of lipids, injection, separation of the esters, identification, and quantitation. The esterification is catalyzed by acids or bases. Split injection or on-column techniques are most often applied. Three types of stationary phases are used in GC separations: very polar, polar, and nonpolar stationary phases (Eder, 1995). The more polar the stationary phase is, the better selectivity the column has with respect to unsaturation of FAs. Nonpolar stationary phases have lower selectivity, but their thermal stability is higher. To get high resolution in separations of a wide range of FA esters, programmed temperature of the capillary column should be used. The easiest way of identification of the esters is comparing their retention times with those of pure standards available commercially. A supplementary method is the calculation of relative retention times or equivalent chain length. Another method of identification is the use of a mass spectrometer coupled on-line with the gas chromatograph. The quantitation of the esters relies on the measurements of the areas of the particular peaks. The areas are proportional to the contents of the esters in the injected sample.

The separations of more complex samples, e.g., industrially hydrogenated fats, may be achieved using silver-ion, high-performance liquid chromatography (HPLC). Silver-ion HPLC separates the FA methyl esters according to unsaturation degree (Christie, 1983). UV-VIS or laser light scattering detectors (LLSDs) are used, but they cause some difficulties in quantitation (Stolyhwo et al., 1983). In the case of a UV-VIS detector, there are different extinction coefficients for different compounds, while the response signal of the LLSD is not linear and the dynamic range is much more limited than that of the flame ionization detector. The use of the flame ionization detector is very limited because of its sensitivity to residues of the organic mobile phase.

COMPOSITION OF TRIACYLGLYCEROLS

Methods of Analysis

TAG analysis provides information on the distribution of FA on the glycerol molecule and on the composition of TAGs in the sample. Thus, the analysis of TAGs in fats and oils is more complicated than FA analysis. This is because of their more complex structure and the possibility of the presence of optical isomers. Several methods can be applied: enzymatic, thin-layer chromatography, HPLC, GC, and supercritical fluid chromatography.

The most recognized method for determination of the composition of TAGs is HPLC analysis. Reversed-phase or silver-ion techniques may be applied to separate the TAGs, although there is no ideal universal detector for HPLC lipid analysis. Quantitation in TAGs analysis is more complex than in GC analysis where the FID detector is successfully applied.

Parallel reversed-phase and silver-ion HPLC separate TAGs according to the partition number and the number of double bonds present in the TAG molecule. HPLC separation combined with enzymatic methods enables stereospecific TAG hydrolysis and the evaluation of TAG composition and structure (Ruiz-Gutierrez and Barron, 1995).

The Composition of Virgin Fats (or Oils)

Olive oil ("Olio extravergine di oliva") and encapsulated evening primrose, pumpkin seed, and borage oil are the most popular virgin oils. All the components of virgin oils are, in general, *cis* unsaturated and satu-

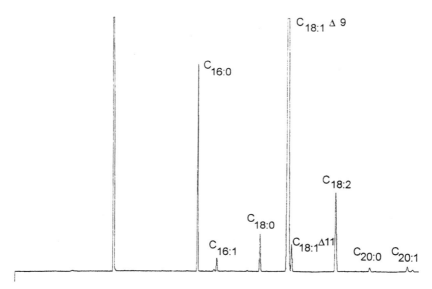

Figure 5.1 Gas chromatogram of fatty acid methyl esters from virgin olive oil ("Olio extravergine di oliva"). Capillary fused silica column, ID 0.25 mm, stationary phase Carbowax 20M, column temperature 175°C.

rated FAs, containing an even number of carbon atoms in the molecule. They do not contain any *trans* isomers. In the case of virgin oils, special care should be taken with regard to the thermal conditions of oil pressing. No component can be added and no component can be removed from the oil during the process. All procedures serve to keep the composition of the virgin oil unchanged, i.e., to preserve the composition exactly the same as in the plant seeds. The composition depends on the plant from which the oil originates. The most frequent components of natural oils are oleic, palmitic, and linoleic acid (Figure 5.1).

The best-known virgin solid fat is butter. It contains short- (from four to fifteen carbon atoms), medium- (from sixteen to twenty), and long-chain (from twenty-one to twenty two) saturated and unsaturated FAs. There are more than seventy different FAs in the butter. The total contents of *trans* FAs isomers does not exceed about 6% and depends on the feeding status of the cow.

The Composition of Refined Oils

The main components of refined oils are the same as those of the respective virgin oils. The principal chemical modification of refined oil is linolenic and linoleic acid isomerization to the highly reactive products that

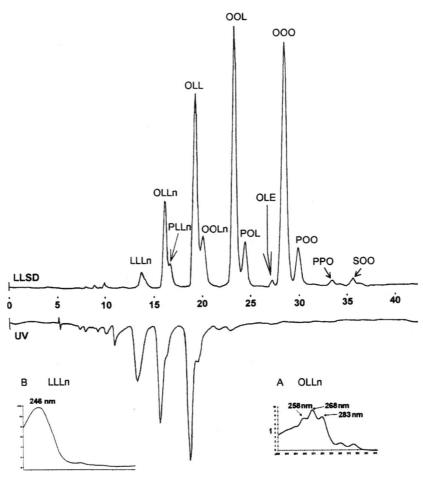

Figure 5.2 High-performance liquid chromatogram of the composition of triacylglycerols from low erucic rapeseed oil recorded using parallel Laser Light Scattering Detector (upper part) and UV-VIS detector (lower part). (A) UV-VIS spectrum of OLLn peak; (B) UV-VIS spectrum of LLLn peak slope. Acids: P−palmitic, S−stearic, O−oleic, L−linoleic, Ln−linolenic, E−erucic. Order of the letters is random and does not mean order of the fatty acid distribution in the particular position in triacylglycerol.

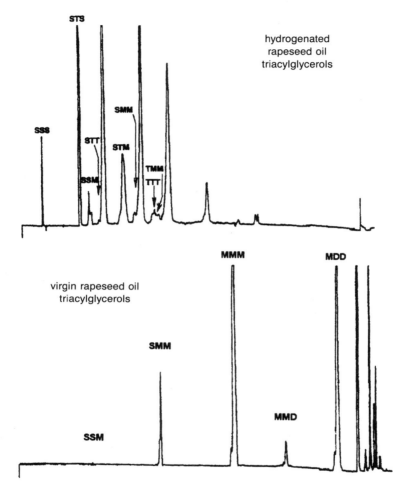

Figure 5.3 HPLC/Ag⁺/LLSD chromatogram of the triacylglycerols from hydrogenated rapeseed oil (part A) and low erucic refined (part B), separation according to the unsaturation degree of the fatty acid carbon chain. Descriptions: fatty acids: S−saturated, T−monounsaturated *trans* configuration, M−monounsaturated *cis* configuration, D−diunsaturated. Separation parameters: Chromspher Lipids column, mobile phase: methylene chloride/acetone/acetonitrile, flow rate 0.8 ml/min, time program composition gradients with different slopes, laser light scattering detector.

have two or three conjugated double bonds in their carbon chain. Groups of conjugated double bonds are easy to detect by UV-VIS detector at wavelengths above 225 nm (Figure 5.2).

The Composition of Hydrogenated Fats

The catalytic fat hydrogenation leads not only to the desired change in the rheological properties of the product, but also to geometrical and positional isomerization of unsaturated FAs. As compared with nonhydrogenated fat, the hydrogenated fat contains new groups of *trans* FA isomers as a result of geometrical isomerization of naturally occurring *cis* isomers. Migration of double bonds across the carbon chain generates the positional isomers of both *cis* and *trans* geometrical FA isomers. For example, from the one naturally occurring monounsaturated *cis* FA isomer having double bond in a precisely defined position at least sixteen different isomers may arise (Stołyhwo, 1995) (Figures 5.3, 5.4).

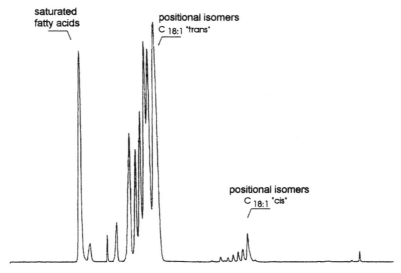

Figure 5.4 HPLC/Ag⁺/LLSD chromatogram of fatty acid methyl esters group composition in frying fat. Separation parameters: Chromspher Lipids column, mobile phase A: hexane/methylene Chloride 90/10 v/v, B: methylene chloride/acetone 10/90 v/v, C: acetone, D: acetone/acetonitrile 20/80 v/v. Gradient elution program: t(min) = 0–100% A; t = 8–88% A/12% B; t = 28–85%A/15% B; t = 36–72% A/28% B; t = 56–52% A/28% B 20% C; t = 70–1% B/98% C/1% D; t = 90–80% C/20% D. Mobile phase flow rate: 0.7 ml/min. Laser light scattering detector.

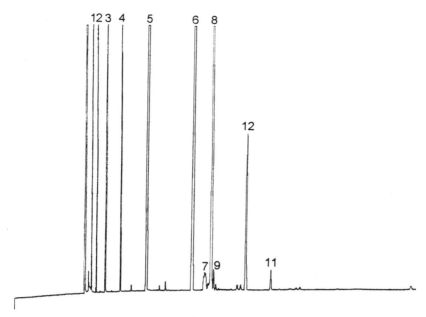

Figure 5.5 Gas chromatogram of fatty acids methyl esters from Hobum transesterified fat. Capillary fused silica column, ID 0.25 mm, stationary phase Carbowax 20M, column temperature 175°C. $1-C_{8:0}$; $2-C_{10:0}$; $3-C_{12:0}$; $4-C_{14:0}$; $5-C_{16:0}$; $6-C_{18:0}$; $7-C_{18:1}$; *trans*; $8-C_{18:1}$ $\Delta 9$; $9-C_{18:1}$ $\Delta 11$; $10-C_{18:2}$; $11-C_{20:0}$.

The Composition of Transesterified Fats

Transesterified fats appear in solid form. They contain short- and medium-chain saturated FAs from eight to eighteen carbon atoms in the molecule and are poor in linoleic acid. The linolenic acid is absent in such fats. The contents of *trans* FA isomers may reach about 1.5% of total FA composition in margarines (Figure 5.5).

CONCLUSIONS

The contemporary analytical methods enable the determination of the chemical effects of industrial refining and modification of fats. Thus, the analytical chemist, armed with such possibilities, may collect the necessary data for medicine and toxicology investigations, which should disclose the influence of the chemically modified fats on human health.

REFERENCES

Christie, W. W., ed. 1988. *HPLC and Lipids*. Oxford, Pergamon Press, pp. 137–144.

Codex Alimentarius Commission, 1983, *Codex Standards for Edible Fats and Oils, CAC, Vol. XI*, Food and Agriculture Organization of the United States of America, World Health Organization, ed. pp. 203–210.

Eder, K. 1995. "Gas chromatographic analysis of fatty acid methyl esters," *J. Chromatogr. B*, 671:113–131.

Ruiz-Gutierrez, V., Barron, L. J. R. 1995. "Methods for the analysis of triacylglycerols," *J. Chromatogr. B*, 671:133–168.

Stołyhwo, A., Colin, H., Guiochon, G. 1983. "Use of light scattering as a detector principle in liquid chromatography," *J. Chromatogr.*, 265:1–18.

Stołyhwo, A., 1995. *Metody analizy żywności* (Methods of food analysis), Merck Conference, Warsaw, pp. 23–39.

CHAPTER 6

Proteins

ZDZISŁAW E. SIKORSKI

THE CHEMICAL STRUCTURE AND PROPERTIES

INTRODUCTION

THE proteins are linear condensation products of different α-L-amino acids (a.a.'s) bound by *trans*-peptide linkages. They differ in the total number of a.a. residues in the molecule and in the number and sequence of each of about twenty various residues along the chain. The chemical properties, size, and distribution of the a.a. affect the conformation of the molecule, i.e., the secondary structure containing helical regions, β-pleated sheets and β-turns, the tertiary structure or the folding of the chain, and the quaternary structure, i.e., the assembly of several polypeptide chains.

The conformation affects the biological activity of the protein, as well as its nutritional value and functional role as a food component.

THE AMINO ACID COMPOSITION

Variation in Composition

In most proteins, the contents of each of the different a.a. residues, calculated as percent of the total number of residues, ranges from 0 to about 30%. In extreme cases, it may even reach 50%. Cereal proteins are generally very poor in lysine. Several major grains are also deficient in threonine, leucine, methionine, valine, and tryptophan. In most collagens, there are no cysteine/cystine and tryptophan residues, and the content of glycine, proline, and alanine is 328, 118, and 104 residues/1,000 residues,

119

respectively. Paramyosin, the myofibrillar protein of marine invertebrates, is characteristic because of amides and other basic residues:
GlN 20–23.5%, AsN 12%, Arg 12%, and Lys 9% (Kantha et al., 1990).
The antifreeze fish serum glycoproteins contain a number of the a.a.
sequence Thr-X_2-Y-X_7, where X is predominantly alanine and Y a polar
residue. Threonine and Y probably form hydrogen bonds with ice crystals,
thereby inhibiting the crystal growth (Price et al., 1990). The β-caseins
contain 14% of proline residues. The molecule has a polar N-terminal
region (1–43) with a charge of -12 and the apolar rest of the molecule
containing most of the proline residues. Such sequence favors the
temperature-, concentration-, and pH-dependent associations of the
protein into threadlike polymers, stabilized mainly by hydrophobic
adherences. The Bowman–Birk trypsin inhibitor consists of seventy-one
a.a. residues in one polypeptide chain rich in loops due to the presence of
seven$-S-S-$bonds. The bovine serum albumin contains one SH group
and seventeen intramolecular disulfide bridges per molecule. Grain prolamines are very rich in glutamine (up to 55%) and proline (up to 30%).
Among the 225 residues of the egg yolk protein phosvitin, there are 122
serine residues.

In most food proteins, however, the variation in the contents of different
a.a.'s is not very high. Generally, the content of acidic residues is the highest and that of histidine, tryptophan, and sulphur-containing a.a.'s is the
lowest. However, the number of residues able to accept a positive charge
is often higher, especially in plant proteins, since about 50% of aspartic
and glutamic acid side-chain carboxyl groups are amidated.

Many a.a. residues undergo enzymatic amidation, hydroxylation, oxidation, esterification, glycosylation, methylation, or cross-linking. Thus, the
total number of different residues in known proteins is about 130. Some
fragments of the polypeptide chains may be removed (Figure 6.1).
Modified residues in a given protein can be used for analytical purposes,
e.g., hydroxyproline, which is characteristic for collagens.

Posttranslational modifications may result in covalent attachment of
nonproteinaceous fragments to the proteins. They may change the ionic
character of the molecule, e.g., the phosphoric acid residues or saccharides. The residues involved in phosphorylation are serine, threonine,
hydroxylysine, histidine, arginine, and lysine. Among the proteins containing many phosphorylated a.a. residues are α_S-caseins. In the central
region of α_{S1}-casein, the phosphorylated serine (PSe) occurs in sequences
PSe-Ala-Glu, PSe-Val-Glu, PSe-Glu-Pse, and PSe-Ile-PSe-PSe-PSe-Glu.
Such composition favors oligomer formation due to hydrophobic interactions of the apolar fragments of the molecules, with the charged sequences
exposed to the solvent. In posttranslational attachment of carbohydrate
moieties the a.a. residues asparagine, serine, threonine, cysteine, hy-

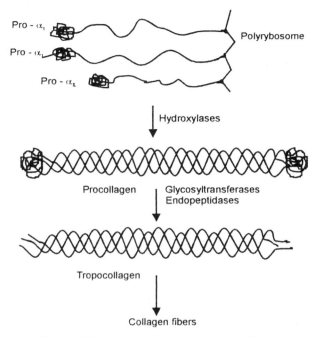

Figure 6.1 Posttranslational modifications in collagen.

droxylysine, and hydroxyproline are involved (Whitaker, 1977). A high content of saccharides is characteristic for the allergenic glycoproteins of soybeans (up to about 40%), several egg white proteins (up to 30%), albumins of cereal grains (up to 15%), whey immunoglobulin (up to 12%), and collagens of marine invertebrates (up to 10%). In $\varkappa$-casein, there is a hydrophobic N-terminal part (1–105) and a hydrophilic macropeptide (106–169), or a glycomacropeptide, with a saccharide moiety (0.5%) of N-acetylneuraminic acid, D-galactose, N-acetylgalactosamine, and D-mannose residues.

Hydrophobicity

Average Hydrophobicity

The nonpolar character of an a.a. can be represented by hydrophobicity, i.e., the change of the free energy F_{ta} accompanying the transfer of the a.a. from water to a less polar solvent:

$$F_{ta} = RT \ln (N_w A_w / N_{org} A_{org}) \tag{6.1}$$

TABLE 6.1. Hydrophobicity of Side Chains
of Amino Acids (water → ethanol).

Amino Acid	F_{tb} (J/mole)	Amino Acid	F_{tb} (J/mole)
Tryptophan	−12,550	Arginine	−3,100
Isoleucine	−12,400	Alanine	−3,100
Tyrosine	−12,000	Glutamic acid	−2,300
Phenylalanine	−11,000	Aspartic acid	−2,250
Proline	−10,850	Histidine	−2,100
Leucine	−10,100	Threonine	−1,850
Valine	−7,050	Serine	−170
Lysine	−6,250	Glycine	0
Methionine	−5,450	Asparagine	40
Cysteine	−4,200	Glutamine	400

where:

N_{org} and N_{org} = solubility in water and organic solvent, respectively, M/dm³

A_{org} and A_{org} = activity coefficients

The hydrophobicity of the side chain of an a.a. (Table 6.1) is given by the formula

$$F_{tr} = T_{ta} - F_{tGly}$$ (6.2)

The average hydrophobicity F_{tav} of a protein can be estimated as

$$F_{tav} = \Sigma F_{ta}/n$$ (6.3)

where:

n = number of a.a. residues in the protein molecule

It is not possible to predict the conformation and behavior of a protein in solution on the basis of F_{tav}. However, proteins of high F_{tav} yield bitter hydrolysates.

Surface Hydrophobicity

A significant impact on the functional properties of proteins is exerted by patches of hydrophobic residues distributed on the surface of the protein molecule, not accommodated in the interior and thus available for interactions with the environment.

The presence of phenylalanine, tyrosine, and tryptophan residues in food proteins can be monitored by measuring the intrinsic fluorescence. These residues absorb ultraviolet radiation and emit fluorescence in the order:

Phenylalanine	260 nm	283 nm
Tyrosine	275 nm	303 nm
Tryptophan	283 nm	343 nm

The intensity of fluorescence and the wavelength of maximum intensity depends upon the polarity of the environment. Thus, the location of a tryptophan residue in a nonpolar region is characterized by emission wavelength 330–332 nm, while at complete exposure to water at 350–353 nm. Furthermore, electron withdrawing groups, like carboxyl, azo, and nitro groups, as well as different salt ions, have a quenching effect on fluorescence. Measurements of intrinsic fluorescence and of fluorescence quenching have not found, however, wide application in hydrophobicity determinations, because they are restricted to the effect of aromatic a.a. residues.

The simplest and most commonly used are hydrophobic probe methods, which are based on the phenomenon that the quantum yield of fluorescence of the compounds containing some conjugated double bond systems is about 100 times higher in a nonpolar environment than in water. Thus, hydrophobic groups can be monitored by using appropriate aromatic or aliphatic probes and fluorescence measurements. Most often used are 1-anilinonaphthalene-8-sulfonate (ANS) (Formula 6.1) and *cis*-parinaric acid (CPA) (Formula 6.2). Also, the binding of triacylglycerols or sodium dodecylsulphate may be determined.

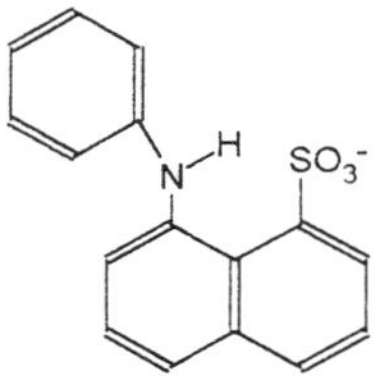

(6.1)

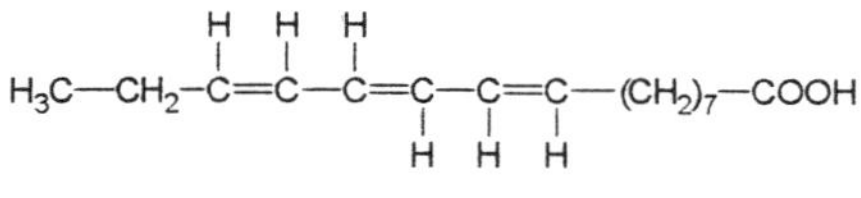

(6.2)

CONFORMATION

The Native State

In a natural environment, the proteins spontaneously fold from an extended form L to the native conformation N, which is affected by the primary structure:

$$L \rightleftharpoons N \qquad (6.4)$$

This is accompanied by a decrease in free energy:

$$-RT \ln K = \Delta G = \Delta H - T\Delta S \qquad (6.5)$$

where

R = gas constant
T = temperature
S = entropy
K = equilibrium constant ($K = [N]/[L]$)

In a biological environment, the conformation of protein molecules is largely affected by the interactions with water. These interactions comprise the formation of hydrogen bonds between water molecules and hydrophilic residues, resulting in enthalpy changes, as well as hydrophobic effects caused by nonpolar residues that bring about entropy changes. This is reflected in the total free energy change ΔG_t:

$$\Delta G_t = \Delta H_p + \Delta H_w - T\Delta S_p - T\Delta S_w \qquad (6.6)$$

where the subscripts p and w refer to protein and water, respectively.

The native protein conformation is stabilized by various forces (Figure 6.2). The dipole–dipole interactions, depolarization, and dispersion forces are significant only at a very close distance r of the atoms because the energy decreases with r^{-6}. The hydrogen bonds, abundant in proteins, differ in energy from about 2 to about 12 kJ/mol, depending on the properties and direction of the groups involved. The strength of the hydrogen bonds does not depend significantly on temperature, but increases with increasing pressure. The energy of the ionic bonds is affected by the dielectric constant and may reach in the hydrophobic core of a protein about 21 kJ/mol between the ionized residues of aspartic acid and lysine. The energy of hydrophobic interactions increases with temperature and decreases with increasing pressure. Covalent bonds other than those in the

Figure 6.2 The structure of actin. [Courtesy of H. G. Mannherz. 1992. "Crystallization of actin in complex with actin-binding proteins," *J. Biol. Chem.*, 267(17):11661–11664. Reprinted with permission of the American Society of Biochemistry & Molecular Biology.]

polypeptide chain, although of highest energy, are generally very limited in number. However, the proteins rich in such bonds have a high thermal stability, e.g., mature collagens containing different cross-links generated in reactions of the oxidized ϵ-NH_2 group of lysine and hydroxylysine and in the Maillard reaction of the saccharide moieties of the molecules, as well as proteins containing many disulfide bridges, e.g., some proteinase inhibitors.

(6.1)

A very significant effect on the functional properties is exerted by the quaternary structures of the proteins.

Soybean glycinin, composed of six basic and six acidic subunits, has a structure of two superimposed rings. In each ring, the three acidic and three basic subunits are arranged alternatively. Thus, electrostatic interactions are possible both within each ring and between the rings. Because the conformation of the oligomer is buttressed by noncovalent forces, addition of urea and changes in pH and ionic strength lead to dissociation of the protein into subunits.

In milk, the caseins are present in free, soluble form and as a colloidal dispersion of micelles. The equilibrium between the soluble and micellar form is affected by temperature, pH, and Ca^{2+}. In fresh milk, about 80–90% of caseins is in the micellar form. The micelles have a diameter of about 100–300 nm, are very porous, and hydrated to about 3.7 g H_2O/g protein. The micelles are formed from several hundreds of subunits, whereby each subunit is composed of twenty-five to thirty monomers. The components are assembled mainly due to hydrophobic associations of apolar fragments. According to different models, the micelles have open structures if composed of subunits containing all kinds of caseins, or else the core of subunits formed from α_S- and β-caseins is covered with a layer of $\varkappa$-caseins. The subunits in the micelles interact with their hydrophobic parts. The main factors responsible for the association are the interactions of polyphosphates, calcium, and citrate with the charged residues in the caseins.

Denaturation

The energy stabilizing the native conformation of proteins is generally rather small. The net thermodynamic stability of the native structure of many proteins is as low as about 40–80 kJ mole^{-1} (Damodaran, 1989). The unfolding enthalpy of cytochrome c, metmyoglobin, and lysozyme is about 210, 285, and 368 kJ mole^{-1}, respectively. Therefore, different factors, e.g., ionizing radiation, shift in pH, change in temperature, concentration of various ions, or addition of detergents or solvents, may dissociate the oligomers into subunits, unfold the tertiary structure, and uncoil the secondary structure (Figure 6.3). These changes are known as denaturation.

Exposure to the environment of the a.a. residues originally buried in the interior of the molecule changes the isoelectric point (pI), surface hydrophobicity, and the biochemical properties of proteins. Denaturation may be reversible, depending on the degree of deconformation and environmental factors. This may affect the validity of determinations of enzyme activity used as a measure of, e.g., severity of heat processing in

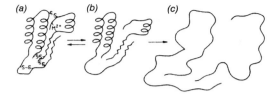

Figure 6.3 Protein denaturation: (a) native molecule, (b) molecule in a changes conformation with ruptured disulfide bridges and ionic bonds, (c) denatured molecule with randomly extended polypeptide chains.

food operations. Recently, the activity determination of the endogenous enzyme γ-glutamyltransferase has been suggested for monitoring milk pasteurization. The enzyme undergoes complete inactivation after 16 s at 77°C and no reactivation has been evidenced (Zehetner et al., 1995).

Generally, the conditions of processing cause irreversible denaturation followed by reactions of the denatured proteins with other food components, which may lead to loss in food quality. However, in foods the protein denaturation may have beneficial or detrimental effects. The main effects comprise changes in pI, hydration, solubility, viscosity of solutions, other functional properties, biological activity, and reactivity of a.a. residues.

THE FUNCTIONAL PROPERTIES

INTRODUCTION

The functional properties of proteins are displayed in interactions with the surrounding solvent, ions, other proteins, saccharides, lipids, and in surface phenomena. The properties most important in food processing can

TABLE 6.2. Functional Properties of Proteins Displayed in Interactions with Different Food Constituents.

Interactions with		
Water	Water and Proteins	Lipids or Gases
Wettability	Viscosity inducing	Emulsifying ability
Swelling	Gelling	Emulsion stabilization
Rehydration	Fiber forming	Foaming ability
Water holding	Dough forming	Foam stabilization
Solubility	Membrane forming	

be roughly grouped accordingly (Table 6.2). They affect the appearance, color, juiciness, mouthfeel and texture of a large variety of foods as well as cutting, mincing, mixing, dough formation, shaping, and transporting of food materials. The functionalities of proteins can actually be modified by using enzymatic and chemical processes. It is hoped that, by better understanding of the tertiary structure of many food proteins, it will also be possible to modify the functionality using genetic engineering (Goodenough, 1995).

SOLUBILITY

Effect of the Protein Structure and Solvent

The solubility depends on the temperature and properties of the protein and of the solvent, pH, concentration, and charge of other ions. Generally, proteins rich in ionizable residues, of low surface hydrophobicity, are soluble in water or dilute salt solutions. Those abundant in hydrophobic groups readily dissolve in organic solvents. The classification of cereal proteins into albumins, globulins, prolamines, and glutelins soluble in water, dilute salt solutions, 60–80% aliphatic alcohols, and 0.2% NaOH, respectively, may be used also for characterization of other proteins. Stabilization by cross-linking is of crucial importance; e.g., the solubility of collagen from different connective tissues depends on the type and age of the tissue. Young tropocollagen can be solubilized in neutral or slightly alkaline NaCl solution; tropocollagen containing intramolecular covalent bonds is soluble in citric acid solution at pH 3, while mature collagen with covalent intermolecular cross-links is not soluble in cold, dilute acids and buffers. It can, however, be partially solubilized in a highly comminuted state or after several hours of treatment in alkaline media.

Denaturation may decrease the protein solubility, e.g., the fish protein concentrate produced by extraction of minced fish with boiling, azeotropic isopropanol is scarcely soluble in water. In organic solvents, due to their low dielectric constant, the energy of interactions between charged residues is higher than in water. This may favor unfolding of the molecules and exposition of the hydrophobic residues, which cannot be counterbalanced by entropy forces. Thermal denaturation, followed by aggregation due to interactions of the exposed reactive groups, leads generally to loss in solubility. On the other hand, if heating brings about deconformation of the quaternary and tertiary structures, it may increase the solubility.

Adding antioxidants to the defatted soy flour prior to alkaline extraction enhances the solubility of the protein isolate in proportion to the decrease in oxidation of thiol groups (Boatright and Hettiarachchy, 1995).

Effect of Ions

In water solutions, the solubility of proteins has a minimum at the pI (Figure 6.4). Because at such pH there is no electrostatic repulsion between the molecules, the hydration layer alone cannot prevent aggregation. Although in the aqueous environment, the attraction of water dipoles by ionized groups of opposite charges in a.a. residues largely offsets the electrostatic binding between the ions, the net balance of the attraction and change in solvent entropy favors salt bridge formation. At pH values more acid or more alkaline than the pI, the solubility increases due to repelling of the positively or negatively charged molecules, as well as due to increased interaction of the charged polypeptide chains with water dipoles. The pI of a protein may shift slightly with changing concentration of salts in the solution.

The effect of ions on the solubility of a protein depends on their ionic strength μ:

$$\mu = 0.5 \ \Sigma m_i Z_i^2 \tag{6.7}$$

where:

m_i = molarity of the solution in respect to the given ion
Z_i = charge of the ion

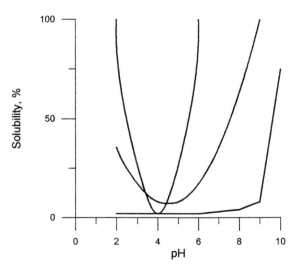

Figure 6.4 The effect of pH on the solubility of proteins.

and on their effect on the surface tension of the solvent, as well as on the dipole moment and the reduction of the molecular surface area of the protein upon aggregation. Various ions, depending on their size and charge favor or decrease the solubility of proteins. In the low range of concentration, i.e., to $\mu = 0.5$–1.0, the solubility increases with increasing concentration of neutral salts. This is known as salting in. The increment of solubility depends on the properties of the respective salt. The ions have a screening effect on the charged protein molecules. Being surrounded by water dipoles, they add to the hydration layer, which favors solubilization of the macromolecules. At higher concentration, the effect depends mainly on the ability of the salts to affect the water structures. Ion–water structures that are more stable than the structures existing in pure water increase the surface tension of the solution and lead to protein precipitation known as salting out. The effect of any given salt on the solubility of proteins depends on the structure-breaking and structure-forming ability of its ions. The ability of ions to affect the salting in of proteins can be represented by the series:

$$SO_4^{2-} < F^- < CH_3COO^- < Cl^- < Br^- < NO_3^- < I^- < ClO_4^-$$

$$< SCN^-, NH_4^+ < K^+ < Na^+ < Li^+ < Mg^{2+} < Ca^{2+}$$

and that of different salts by the order of their increments of surface tension in solutions:

$$KSCN < KI < KBr < NaCl < Na_2HPO_4 < MgSO_4$$

$$< (NH_4)_2SO_4 < BaCl_2 < MgCl_2 < CaCl_2$$

The ions/salts on the left belong to the salting-in type, those on the right to the salting-out type.

Importance in Food Processing

The solubility versus pH curve is the basis for selecting parameters for extraction of proteins from plant material. Adding a required amount of salt to the meats during cutting and mixing in a silent cutter is a prerequisite for extracting myofibrillar proteins from the tissue structures and to form a sausage batter of proper quality. $CaCl_2$ is used to precipitate the whey proteins, while $CaSO_4$ coagulates soy proteins in tofu manufacturing.

The solubility also contributes to other functional properties of proteins, e.g., gelling and emulsifying properties. It may also be a prerequisite for efficient use of various protein isolates as functional food additives in formulations differing in pH and salt content. The loss in solubility due to abuse treatment of the materials is often indicative of protein denaturation and subsequent cross-linking; therefore, solubility data, if used to characterize commercial products, should be determined in standardized procedures. In the methodology of solubility determination, the following factors must be considered: size and disintegration of the sample, pH and ionic strength of the solvent, proportion of sample size to that of the solvent, number of extractions, foaming during blending and stirring, temperature and time of extraction, and separation of nonproteinaceous material.

WATER-HOLDING CAPACITY

Effect of the State of the Proteins and Tissues

The ability of proteinaceous products to retain water is mainly affected by the protein structures. In meat and fish tissues, the state of water depends on various interactions of water structures with proteins and other solutes. Furthermore, because of the fibrous nature of the muscle and compartmentation caused by different membranes, water is also held in the meat by physical entrapment. Alterations in the spatial arrangement of the proteins and tissue structures caused by biochemical and processing factors are responsible for shrinking or swelling of the material and thus for retaining or exudation of water. Classical works on the effect of pH, divalent cations, postmortem changes, freezing and thawing, heating, salting, polyphosphates, and citrates on the water-holding capacity (WHC) of meat were published by Hamm (1960, 1970, 1986).

GELLING

The Gel Structure

A gel consists of a three-dimensional lattice of large molecules or aggregates capable of immobilizing solvent, solutes, and filling material. Food gels may be formed by polysaccharides, proteins, or both kinds of polymers. The characteristic features of the network are junction zones, where the polymers interact, as well as large fragments, where the macromolecules are randomly extended. The lattice is responsible for the elasticity and the textural strength of the gel. In multicomponent gels, all components may actively participate in the structure by forming separate or coupled networks, or else one category of the polymer, not involved in network formation, may indirectly affect the gelling by steric exclusion of the active molecules. Such exclusion leads to an increase in the concentration of the active component in the volume of the solution where the gel is formed (Oakenfull, 1987). Lipid-filled milk protein gels containing small fat globules with a narrow particle size distribution have smooth texture and high shear modulus.

Generally, gelation is a two-step phenomenon. The first step usually involves dissociation of the quaternary structure of the protein, followed by unfolding of the molecules. In some proteins, heating to about 40°C is sufficient, and some fish protein sols turn slowly into gels even at 4°C. In ovalbumin solutions, gelling starts at 61–70°C. In the second step, at higher temperature, the unfolded molecules rearrange and interact, usually with their hydrophobic fragments, forming the lattice. Ovalbumin gels increase in firmness when heated up to about 85°C. Subsequent cooling generally stabilizes the gel structure. If the rate of the structuring stage is lower than that of denaturation, the unfolded protein molecules can rearrange and interact to form an ordered lattice of a heat reversible, translucent gel. If the interactions in the denatured state occur too rapidly, the molecules form an irreversible coagulum due to random associations to insoluble, large-molecular-weight polymers.

Factors Affecting Gelling

The structure formed by heated protein-water systems depends upon the proteins, the environmental factors, and the process parameters. The coagulum-type proteins generally have molecular weight over 60 kDa and contain over 30% hydrophobic residues, e.g., hemoglobin and egg white albumin. The gelling-type proteins contain less hydrophobic residues and are represented by some soybean proteins, ovomucoid, and gelatin.

The gel strength of heat-set globular proteins at a given concentration is primarily related to the size and shape of the polypeptides in the gel network, i.e., to the weight-average molecular weight. Globular proteins of molecular weight lower than 23,000 do not form gels under normal conditions (Wang and Damodaran, 1990).

In food products, the interaction of different proteins may decrease the gel strength, may have no influence on the rheological properties of the gel, or may be reflected in a synergistic effect. The heat gelation of whey proteins may deteriorate by the presence of small amounts of caseinates. On the other hand, casein micelles in the whey protein matrix may enhance or decrease the gelling, depending, e.g., on pH. The gel strength of kamaboko made of pelagic fish is usually low. This is partly due to the fact that heat coagulation of the sarcoplasmic proteins impair the gelation of actomyosin. In many foods, special factors may significantly affect the gels, e.g., the proteinase catalyzed softening known as modori in minced fish products. This softening may be decreased by adding protease inhibitors from potato, bovine plasma, or egg white. The effect of other factors may be controlled by applying optimum processing parameters.

Gels may be formed in protein solutions and in dispersions, micelles, or even in partially disrupted tissue structures, like in meat and fish products. At high protein concentration, gelation is less influenced by the environmental conditions, especially by pH and ionic strength.

Binding Forces and Process Factors

The hydrophobic interactions prevail at higher temperature and probably initiate the lattice formation, while hydrogen bonds increase the stability of the cooled system. The electrostatic interactions depend upon the pH, charge of the molecules, ionic strength, and the presence of divalent ions. Intermolecular disulfide bridges may also add to the gel formation.

Depending on the properties and concentration of the protein, ionic strength, and pH, even a heat-set gel like that of ovalbumin can be melted by repeated heating and set again when cooled (Shinizu et al., 1991). Heat-induced gels may melt under increased pressure at room temperature, while cold-set gels of gelatin are resistant to such conditions (Doi et al., 1991). Optimum ionic strength and concentration of Ca^{2+} is required for producing well-hydrated heat-set gels from whey proteins.

There is generally a pH range at which the gel strength in the given system is the highest. It depends on the nature of the polymers participating in cross-linking and increases with protein concentration. At the pI of the proteins, due to lack of electrostatic repulsion, the rate of aggregation is usually high, leading to less ordered, less expanded, and less hydrated gels. Well-hydrated gels contain up to 98% water.

The ovalbumin gel has optimum rheological properties at pH 9, while at pH < 6, it is brittle and has low elasticity. Transparent ovalbumin gel can be made by heating at a pH other than pI at a certain salt concentration. In a two-step procedure, transparent gels can be made from ovalbumin, bovine serum albumin, and lysozyme solutions over a broad range of salt concentration by heating the protein solutions first without salt and after cooling by repeated heating in the presence of added salt (Tani et al., 1993). The pH range for gelation of whey proteins is 2.5–9.5, although near the pI, which is about 5, the gels are opaque and coarse and may turn into curd-like coagulum. In the neutral to alkaline pH, the gels made of fat-free whey protein isolates or purified β-lactoglobulin are translucent, smooth, and elastic. In acid conditions, if high shear force is applied for a short time at denaturation temperature, aggregation of whey proteins to microparticles occurs. This leads to well-hydrated gels of a smooth, nonelastic texture, similar to that of a fat emulsion. The rheological properties of whey protein gels, at different pH, depend also on the concentration of Ca^{2+}. The factors affecting gelation of whey proteins were discussed by Mulvihill and Kinsella (1987).

The S−S bridges are responsible for the thermal stability of gels. Such bonds add to the elasticity of heat-set whey protein gels at neutral to alkaline pH but not in acidic conditions when the thiol has low reactivity (Jost, 1993). Ascorbic acid improves the formation of heat-set gels of ovalbumin and kamaboko by undergoing rapid oxidation to dehydroascorbic acid, which affects protein polymerization by intermolecular disulfide bridges (Nishimura et al., 1992). According to Wang and Damodaran (1990), the role of cysteine–cystine interactions in gelation is related to the polypeptide size-increasing effect, rather than to the formation of specific crosslinks in the network.

Gels stabilized mainly at low temperature by hydrogen bonding are heat reversible; i.e., they melt due to heating and can be set again by cooling, while gels stabilized by hydrophobic interactions and covalent bonds are heat stable.

To evaluate the protein functionality in different systems, the quantitative structure-activity relationship approach may be applied (Nakai and Li-Chan, 1988). The gel strength of a meat mince can be predicted using the following formula:

$$G = -0.029ANS - 0.38F - 3.15\,pH + 0.17D$$

$$+ 0.000024ANS^2 - 2.70SEP^2 - 0.0018W^2 \qquad (6.8)$$

$$+ 0.0085S \times SH + 0.44SEP \times P + 27.24$$

where:

G = gel strength, N
ANS = hydrophobicity determined by ANS probe
F = fat content, %
D = dispersibility, %, in salt solution after 1,100 × g, 10 min
SEP = salt extractable protein, %
S = solubility, %, in salt solution after 27,000 × g, 30 min
W = moisture content, %
SH = thiol group content, $\mu M/g$
P = protein content, %

Importance in Food Processing

Gelling is of utmost importance for the quality of comminuted-type, cooked sausages and of kamaboko or similar, traditional Japanese com-

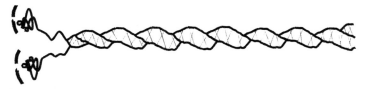

Figure 6.5 A schematic representation of the myosin molecule.

modities made of surimi. The gel strength of such products is mainly affected by the properties of myosin and processing conditions. Comminution of the meat with salt brings about unfolding of the myosin microfibrils and increase of the surface hydrophobicity, resulting in hydrophobic associations in the lattice structure. Heating to 50–80°C favors deconformation of the myosin heads (Figure 6.5) and their interactions. Although myosin has the highest gel-forming ability of all muscle proteins, the whole myofibrillar protein fraction, the sarcoplasmic, and the connective tissue proteins are also capable of gelation. The overall gel strength is not an additive function. It depends on the concentration and interactions of different protein fractions (Lan et al., 1995).

EMULSIFYING PROPERTIES

The Mode of Action

Food proteins help to form and stabilize emulsions, i.e., dispersions of small liquid droplets in the continuous phase of an immiscible liquid. The decrease in the diameter of the droplets due to emulsification exponentially increases the interfacial area. The work (W) required for the increase in surface area (ΔA) can be decreased by lowering the surface tension (z):

$$W = z\Delta A \qquad (6.9)$$

due to attachment of proteins to the droplets. The protein film around the lipid globules, with its electrostatic charge and steric hindrance, prevents flocculation, i.e., formation of clusters of globules and thus more rapid creaming due to the action of gravitational force:

$$V = 2r^2 \, g \, \Delta P/9 \, \mu \qquad (6.10)$$

where:

$V = $ velocity of the droplet

g = gravitational force
ΔP = difference in density of both phases
μ = viscosity of the continuous phase
r = radius of the droplet or cluster of droplets

Stable films around the fat globules also prevent coalescence of the dispersed phase, i.e., joining of the fat globules to form a continuous phase. Furthermore, soluble proteins increase the viscosity of the dispersing phase, thus reducing the rate of creaming and coalescence.

The efficiency of proteins as emulsifiers depends upon their surface hydrophobicity and charge, steric effects, elasticity-rigidity, and viscosity in solution. Globular proteins, which have stable structures and are very hydrophilic, are good emulsifiers only when unfolded. However, the emulsifying properties do not increase linearly with the hydrophobicity of the protein, because they depend on the hydrophile/lipophile balance (HLB), which is defined as:

$$\text{HLB} = 20 \, W_h / W_t \qquad (6.11)$$

where:

W_h = weight of hydrophilic groups
W_t = total weight of the molecule

The emulsifiers with HLB < 9 are regarded as hydrophobic, HLB 11–20 as hydrophilic, and HLB 8–11 as intermediate. There is an effect of protein solubility because the molecules must be able to migrate to the surface of the fat globules. However, in comminuted sausage batters in the presence of salt, the insoluble proteins also may participate in the formation of fat dispersions. After a few minutes of homogenization, about 90% of initially insoluble meat proteins of the stroma can be found in the emulsion layer (Nakai and Li-Chan, 1988).

The quantity of protein required for stabilization of an emulsion increases with the volume of the dispersed phase and with the decrease in diameter of the droplets. The concentration of proteins forming at the interface of a monomolecular layer is of the order of 0.1 mg/m^2, and the effective concentrations are in the range of 0.5–20 mg/m^2. For a high rate of film formation, the concentration of protein in the emulsion may be as high as 0.5 to 5%.

Factors Affecting Emulsifying

The pH of the environment affects the emulsifying properties by changing the solubility of proteins, the surface hydrophobicity, and the charge of the protective layer around the lipid globules. Ions alter the electrostatic interactions, conformation, and the solubility of the proteins. However, in many foods, mainly comminuted meat batters, the concentration of NaCl is considerably high for sensory reasons, so that small changes in the salt content within the accepted range may have no significant effect on the properties of proteins. Heating to about 40–60°C may cause a partial unfolding of the protein structure without loss in solubility and may induce gelation of the protective layer, as well as decrease the viscosity of the continuous phase. Thus, moderate heating may improve the emulsifying properties of proteins.

Determination of Emulsifying Properties

Several procedures are used to determine the efficiency of proteins in emulsifying lipids and the stability that the proteins impart to the emulsions.

The emulsifying capacity (EC) is represented by the volume of oil (cm^3) that is emulsified in a model system by 1 g of protein when oil is added continuously to a stirred aliquot of solution or dispersion of the tested protein. EC is determined by measuring the quantity of oil at the point of phase inversion. The latter can be detected by a change in color, viscosity, or electrical resistance of the emulsion, or the power taken by the stirrer engine. The EC decreases with increasing concentration of protein in the aqueous volume. It is affected by the parameters of emulsification, depending on the equipment, as well as by the properties of the oil.

The emulsion stability (ES) is measured as the final volume of the emulsion V_1 after centrifuging the initial volume V_0 or standing for several hours at specified conditions:

$$ES = (V_1/V_0) \cdot 100 \qquad (6.12)$$

It may also be determined as the quantity of oil or cream separated from the emulsion or the time required for the emulsion to release a specified quantity of oil.

For predicting the emulsifying properties of samples of beef, pork, chicken, and fish, the following formula has been proposed (Nakai and Li-Chan, 1988):

$$EC = 0.091D - 0.024CPA - 0.00046ANS^2 + 0.0015W^2$$

$$+ 0.00028S \cdot CPA + 0.0004ANS \cdot CPA + 33.26 \qquad (6.13)$$

where CPA is the hydrophobicity determined by CPA probe.

FOAMING PROPERTIES

Food foams are dispersions of gas bubbles in a continuous liquid or semisolid phase. Foaming is responsible for the desirable rheological properties of many foods, e.g., the texture of bread, cakes, whipped cream, ice cream, beer froth, etc. Thus, foam stability may be an important food quality criterion. However, foams are often a nuisance for the food processor, e.g., in the production of potato starch or sugar and in generation of yeast. Residues of antifoaming aids in molasses may drastically reduce the yield in citric acid fermentation.

The gas bubbles in food foams are separated by sheets of the continuous phase, composed of two films of proteins adsorbed on the interface between a pair of gas bubbles, with a thin layer of liquid in between. The volume of the gas bubble may make up 99% of the total foam volume. The contents of protein in foamed products is 0.1–10% and of the order of 1 mg/m² interface. The system is stabilized by lowering the gas–liquid interfacial tension and formation of rupture-resistant, elastic protein film surrounding the bubbles, as well as by the viscosity of the liquid phase. The foams, if not fixed by heat setting of the protein network, may be destabilized by drainage of the liquid from the intersheet space due to gravity, pressure, or evaporation; by diffusion of the gas from the smaller to the larger bubbles; or by the coalescence of the bubbles resulting from rupture of the protein films.

Factors facilitating the migration of the protein to the interface and formation of the appropriate film are important for foaming. The foaming capacity or foaming power, i.e., the ability to promote foaming of a system, measured by the increase in volume, is affected mainly by the surface hydrophobicity of the protein. The stability and strength of the foam, measured by the rate of drainage and the resistance to compression, respectively, depends on the flexibility and the rheological properties of the protein film. Other components of the system, mainly salts, sugars, and lipids, affect the foam formation and stability by either changing the properties of the proteins or the viscosity of the continuous phase. The standard whey protein concentrates, which contain 4–7% of residual milk lipids, have significantly inferior foaming properties than lipid-free isolates.

Excellent foaming properties are characteristic for the egg white proteins, especially ovalbumin, ovotransferrin, and ovomucoid (Trziszka, 1994). Chilling of the egg white below room temperature or the presence of sugar or lipids decreases the foaming, pH < 6 increases the foaming capacity, while heating of the dried proteins at 80°C for a few days before use increases the foam stability. The foaming properties of whey protein isolate can be improved by addition of Ca^{2+} or Mg^{2+}. The ions are effective only immediately after the salts are added. According to Zhu and Damodaran (1994), the ions might cause unfolding and polymerization of the proteins at the interface via ionic linkages. On the other hand, prolonged incubation of the isolate solution with the salts slightly reduces the film-forming ability, possibly by promoting aggregation and micellization of the protein.

CHANGES DUE TO PROCESSING

HEATING CHANGES

Introduction

Heat, as applied in food processing and cookery, may induce desirable and undesirable reactions in proteins. They comprise denaturation, loss in enzyme activity, change in solubility and hydration, discolorations, derivatization of a.a. residues, cross-linking, rupturing of peptide bonds, and formation of sensory active compounds. These processes are affected by the temperature and time of heating, pH, oxidizing compounds, antioxidants, radicals, and other reactive constituents, especially reducing saccharides. Some of the undesirable reactions can be minimized. Stabilizers, e.g., polyphosphates and citrate, which bind Ca^{2+}, increase the heat stability of whey proteins at neutral pH. Lactose present in whey in sufficiently high concentration may protect the proteins from denaturation during spray drying (Jost, 1993).

The susceptibility of proteins to thermal denaturation depends on their structure, mainly on the number of cross-links, as well as on simultaneous action of other denaturing agents. In many proteins, the first changes appear at 35–40°C. The Q_{10} for denaturation in the range of maximum change may be as high as 100 to 600.

Heating of soluble collagen brings about uncoiling of the superhelix and unfolding and separation from each other of polypeptide chains if they are not connected by stable bonds. This results in a decrease in viscosity of the solution. The denaturation temperature T_d can be measured in the range of rapid changes in viscosity. Insoluble collagen fibers can also be heat denatured. This leads to shrinking of the fiber by up to 75% of the original length. The shrinkage temperature T_s is usually about 20°C higher than T_d. It increases with cross-linking and in fish collagen with the contents of hydroxyproline. Thermally denatured collagen turns into gelatin after prolonged boiling in water, due to partial hydrolysis. This phenomenon affects the rheological properties of cooked meats containing much connective tissue.

Rheological Changes

Proteins are primarily responsible for the rheological properties of meat, poultry, fish, cheese, bread, cakes, and many other foods.
The texture of muscle foods is affected mainly by

- the contents and cross-linking of collagen and the morphological structure of the tissues, e.g., meat, fish, and squid
- the biochemical state of the muscle, i.e., interactions of the proteins of the myofibril, e.g., in rigor mortis, cold shortening, thaw rigor, as well as the proteolytic changes during aging of the meat
- mechanical disintegration of the muscle structure

Abuse treatment of food during storage, processing, and preparation may lead to deterioration in rheological properties of many products, e.g., toughening and partial loss of gel-forming ability of frozen stored fish, shrinking and toughening of pasteurized ham, toughening and formation of grainy structure in casein curd, and separation of fat layers in sausages.

Thermal changes comprise loss in solubility as the result of aggregation of molecules, e.g., the whey proteins β-lactoglobulin and the immuno-globulins, or increase in solubility due to decomposition of superstruc-tures like in collagen, formation of gels, e.g., in most types of sausages, kamaboko, meat gels, and several types of cheeses, development of gas re-taining structures as in the dough and bread, hydrolytic changes, and alteration of the rate of proteolysis.

The thermal changes in meat commence at about 40°C – some of the meat proteins in solution coagulate at that temperature. Further heating results in shrinkage of collagen, at 50–60°C, followed by gelatinization in a moist environment. At about 65°C, hardening of the myofibrils occurs. The final texture of the product depends upon hardening of the myofibrillar structure and gelatinization of collagen in the given meat at the particular state of post-mortem changes, heated according to a particular time–temperature regime (Figure 6.6). The rheological changes in meat and fish are emphasized by drip loss, 20–40% of the original weight, and by shrinkage by up to about 25%.

In making bread, the wheat flour, when mixed with water, forms a vis-coelastic dough that retains gas and sets due to heating during baking. The agent responsible for these dough properties is gluten, which develops in mixing of the flour with water. Dry gluten contains 80–90% proteins, 5–10% saccharides, 5–10% lipids, and some minerals. The gluten proteins are composed of 40–50% gliadin, 35–40% glutenin, and 3–7% soluble proteins. During mixing of flour with water, the flour particles, composed of starch granules embedded in the protein matrix, are progressively hydrated and finally form the viscoelastic dough. SH/S–S interchange

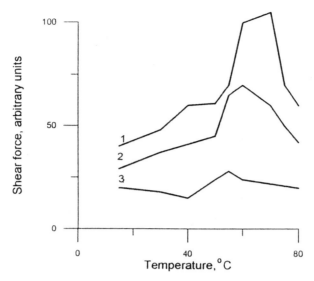

Figure 6.6 The effect of temperature and pH on the rheological properties of fish meat. (Courtesy of E. Dunajski.)

reactions lead to the formation of intermolecular disulfide bridges. Generally, the resistance to extension of the dough increases to a peak value. Overmixing usually results in a decrease in resistance. The effects of overmixing are caused by excessive breaking of the S−S bonds, not accompanied by formation of new intermolecular bridges. This results in depolymerization of the protein and, thus, increased solubility and decreased viscosity of the dough. The required mixing time can be controlled, within limits, by changing the pH, adding oxidants, or mixing in an inert atmosphere. The gas-retaining ability results from high viscosity of the hydrated protein-carbohydrate-lipid system. The flours made of rice and corn do not decrease the gas diffusion rate as effectively, because the doughs prepared from these flours are not as viscoelastic as the gluten-containing dough. The relatively high gas retention of rye flour doughs is due to the viscosity of soluble pentosans. A similar effect can be achieved in gluten-free doughs by adding some surfactants, particularly natural flour lipids, monoacylglycerols, or xanthan gum (Hoseney and Rogers, 1990). During baking, the expansion of the dough stops due to polymerization and/or due to cross-linking as a result of SH/S−S interchange reactions. The setting prevents further expansion of the dough and the collapse of the loaf of bread.

Very severe heating of foods, at much higher temperature and time than required for sterilization, may lead to formation of isopeptide cross-links between the free NH_2 group of lysine and the carboxylic group of aspartic or glutamic acids:

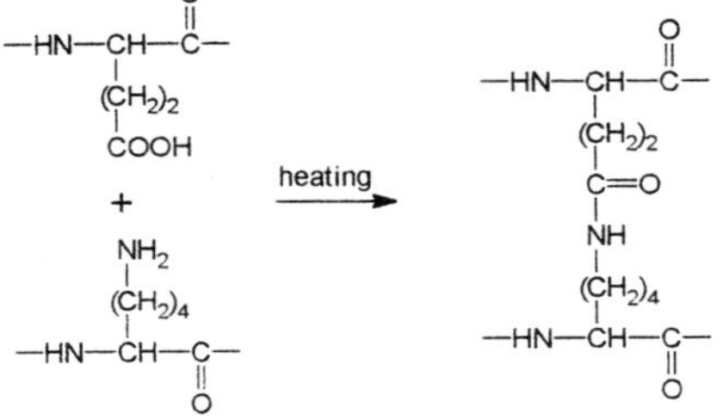

(6.2)

ε - N (γ - glutamyl) lysylamide crosslink

These reactions tighten the structure of the products and may decrease the biological availability of lysine, as well as the digestibility of the proteins.

Changes in Color

The rheological changes in cooked meats are accompanied by alterations in myoglobin, hemoglobin, and cytochromes. Pig, sheep, beef, and wheat contain 0.2–2.4, 4.5–5.5, 1–20, and 50 mg of myoglobin in 1 g of muscle, respectively. The total content of all chromoproteins in the red muscles of fish range from a few to 20 mg/g and is about twenty times higher than in the white muscles. Changes in the chromoproteins due to oxidation/reduction, denaturation, curing, and reactions with sulphur-containing compounds bring about desirable or undesirable alterations in color (see Chapter 8).

Beef heated to 58–60°C internal temperature is rare, to 66–68°C medium rare, to 73–75°C medium, and 80–82°C well done. Recommended endpoint temperature in pork is 77°C and in poultry 77–82°C.

The reactions of myoglobin in cured, heated meat, leading to the formation of the heat-stable nitrosylhemochromogen and a nitrite-protein complex, proceed according to Killday et al. (1988) as follows:

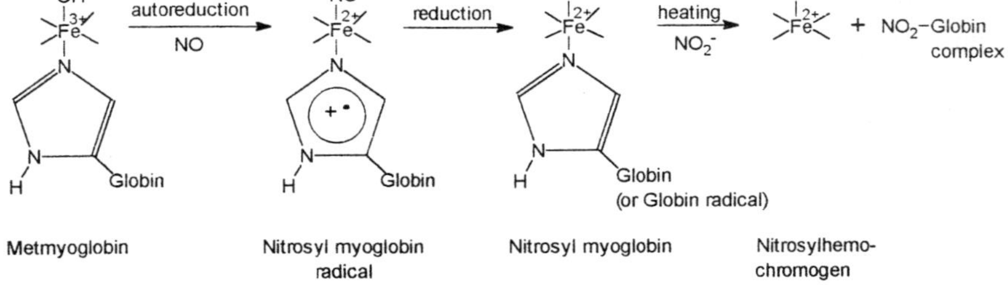

(6.3)

In proteinaceous foods rich in carbohydrates or secondary lipid oxidation products, the Maillard reaction prevails. The generation of pyrraline from lysine can be used as an indicator of thermal changes of proteins.

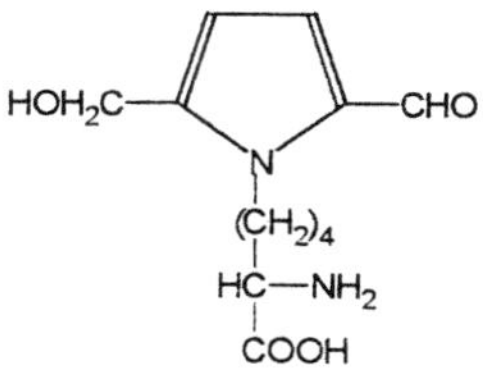

Pyrraline

(6.3)

Development of Volatile Flavor Compounds

Severe heating of proteinaceous foods leads not only to development of flavor compounds due to the Maillard reaction, but also to thermal degradation of methionine, cysteine, and cystine residues in proteins, as well as of different low-molecular-weight compounds. These reactions are discussed in Chapter 9.

Reactions at Alkaline pH

Alkaline treatment is used for peeling of fruits and vegetables, for producing protein isolates, for removing nucleic acids from single-cell protein preparations, and for inactivating mycotoxins and proteinase inhibitors. In protein solutions and in foods, severe changes in reactive a.a. residues take place at high pH, even at temperatures as low as 50°C. They lead to cross-linking and formation of nontypical a.a. residues.

Cross-linking is based on β-elimination, mainly in cysteine, serine, phosphoserine, or threonine, followed by a nucleophilic addition of the ϵ-NH$_2$ of lysine, δ-NH$_2$ of ornithine, SH of cysteine residues, or of ammonia to the double bond of dehydroalanine or 3-methyldehydroalanine:

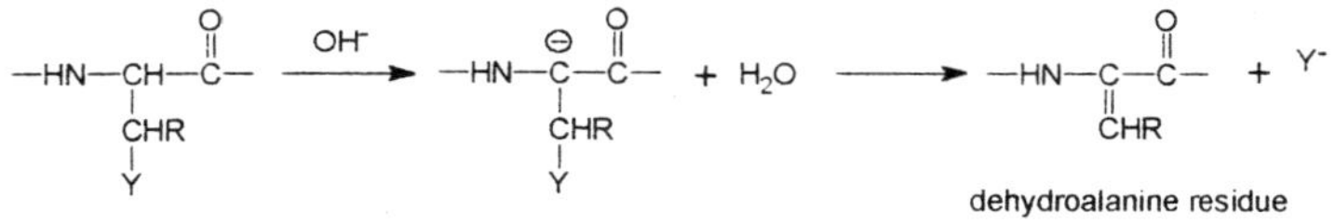

dehydroalanine residue

where: R=H, CH$_3$; Y=OH, OPO$_3$H$_2$, SR$^+$, SSR

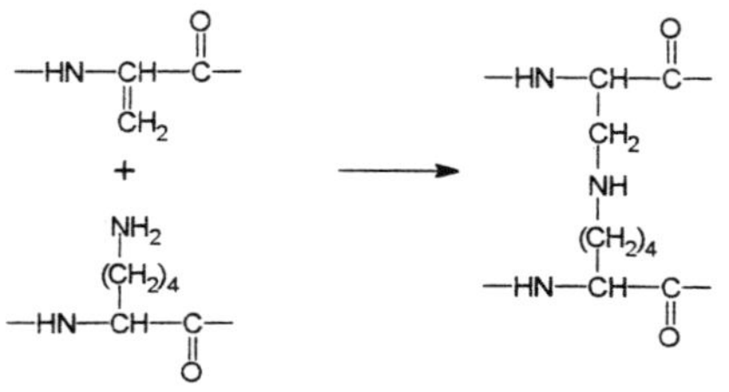

(6.4)

Reactions of cystine at alkaline pH also liberate free sulphur and a sulfide ion:

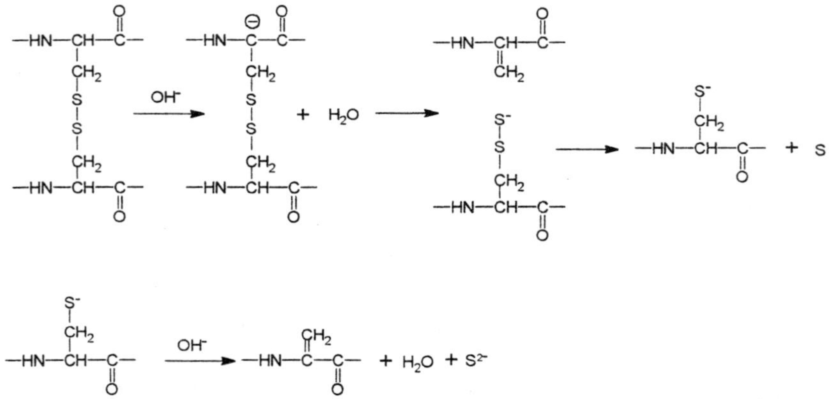

(6.5)

At a given temperature, the overall rate of reaction depends upon the rate of β-elimination and on the conformation of the protein. This is because the accessibility of the dehydroalanine residue for the nucleophilic attack depends on the spatial arrangement of the reacting groups. The reaction can be inhibited by acylating the nucleophilic groups in proteins or by adding thiol compounds, which compete with a.a. residues for the dehydroalanine double bond:

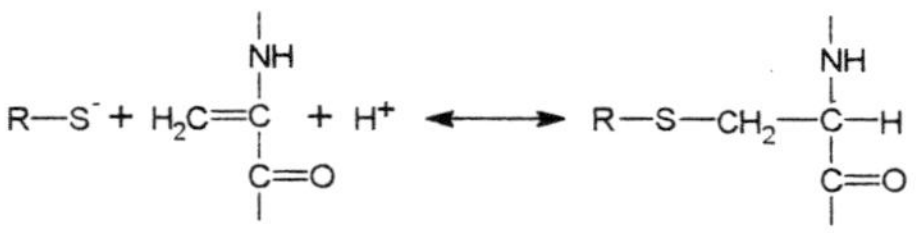

(6.6)

Alkaline conditions favoring cross-linking in proteins also lead to racemization of a.a.'s:

$$\text{L-amino acid} \rightleftharpoons \text{D-amino acid}$$

i.e., after the first step of the reaction sequence, the carboanions recombine with protons to the L- and D-forms. The rate of racemization is affected mainly by the properties of the residues – the lowest is in aliphatic a.a. – and on the structure of the protein. Generally, the rate of the process is about ten times higher in proteins than in free a.a.'s.

Severe heating at alkaline pH may decrease the digestibility and biological value of proteins. Loss in digestibility results from cross-linking and racemization. The rate of absorption of some D-a.a.'s in the organism is lower than that of corresponding L-forms. Not all D-a.a.'s can be metabolized. Furthermore, some of the modified residues, mainly lysinoalanine, induce pathological kidney changes in experimental animals.

OXIDATIVE CHANGES

In most proteinaceous foods, the oxidation of a.a. residues in proteins is initiated by radicals and different reactive forms of oxygen generated by radiation, as well as by lipid peroxides and other oxidation products. Polyphenols, present in many foods, are prone to oxidation to quinones by oxygen at neutral and alkaline pH and can also act as strong oxidizing agents in different products. Also, H_2O_2, if abused as a bactericidal agent, e.g., in treatment of storage tanks, packaging materials, or proteinaceous meals may cause oxidation of proteins.

The effect of oxidative changes in proteins depends upon the activity of the oxidizing agent, the presence of inhibitors and sensitizers, e.g., riboflavin, the temperature, and the reactivity of the a.a. residues. Muscle tissues contain several prooxidants capable of generating radicals, e.g., transition metals, heme pigments, reducing compounds, mitochondria, microsomes, and enzymes. Enzyme-generated hydroxyl radical HO⋅ can readily react with proteins in muscle systems. Abstraction of a hydrogen atom on the α-carbon or in the a.a. side chain of a protein leads to the formation of protein radicals. These may polymerize with other protein or lipid radicals. Also, different scission products may be formed. Hydrogen sulfide and free sulphur may be generated due to oxidation of the thiol group. The sulphur-containing a.a.'s are converted to many oxidized compounds, including cysteine sulfenic, sulfinic and sulfonic acids, mono- and disulfoxides, and mono- and dissulfones. The a.a.'s most prone to oxidation in different model systems and in foods are methionine:

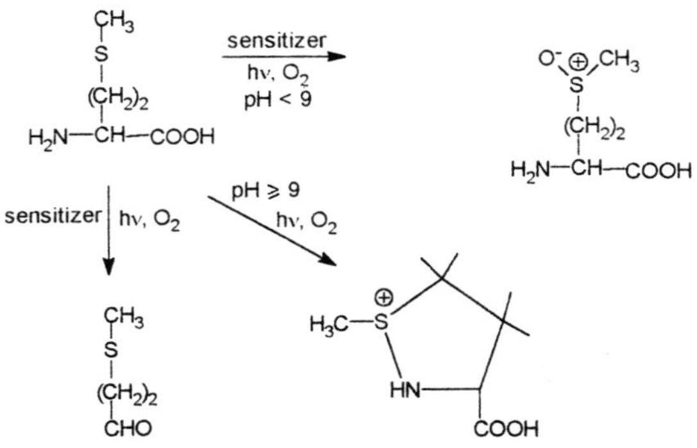

(6.7)

followed by cysteine, tryptophan:

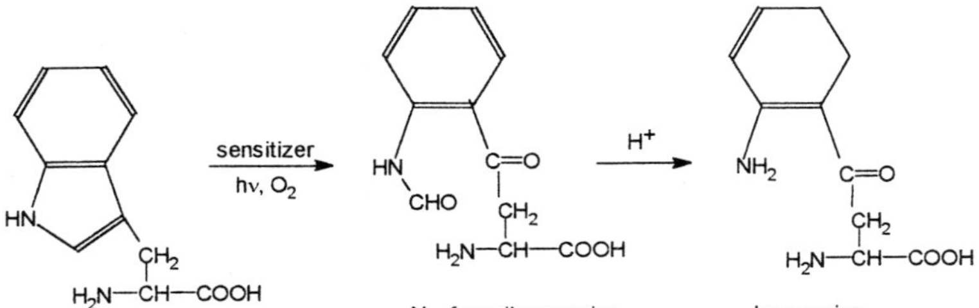

N - formylkynurenine   kynurenine

(6.8)

tyrosine, and histidine:

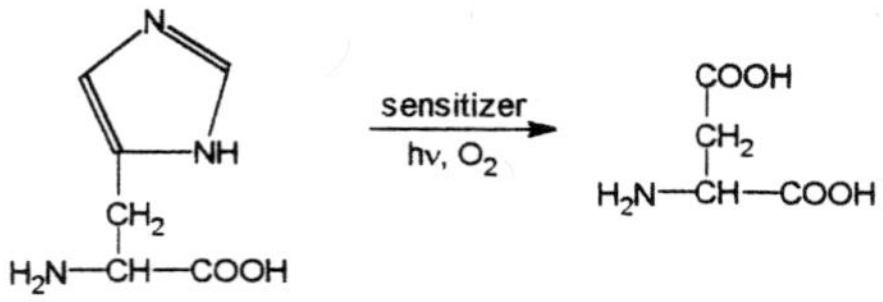

(6.9)

The products of oxidation affect the flavor of foods, either directly, or by reacting with precursors. Their biological value depends generally upon the degree of change due to scission, polymerization, and oxidation. Furthermore, due to oxidation, the contents of essential a.a.'s in foods may be reduced so much that they may become limiting in the diet. Formation of protein–protein and protein–lipid cross-links may decrease the digestibility of proteins.

ENZYMATIC CHANGES

CHANGES IN PROTEINS IN SITU

Some proteins affect the sensory and functional properties of foods by exhibiting enzymatic activity in their natural environment or when added intentionally during processing in the form of enzyme-rich materials, pure enzyme preparations, or starter cultures of microorganisms. The examples of effects involving protein changes are loss of prime freshness in fish after catch, rigor mortis and tenderization in meat, ripening of salted fish and cheese, softening of fish gels, fermentation in soybeans processing, and proteolytic changes in wheat flour.

In milk, the phosphatases may dephosphorylate the caseins. Thiol oxidase may participate in oxidation of SH groups to disulfides, while plasmin, the protease associated with casein micelles, attacks β-casein and α_s-caseins. The caseins are also hydrolyzed by bacterial proteases present in the milk.

The sensory quality of meat and seafoods is significantly affected by the endogenous proteases. The lysosomes contain, among other enzymes, at least twelve cathepsins, which can exert a concerted hydrolytic action on proteins and peptides. The optimum pH for the activity of cathepsins is generally in a low acidic range, although many enzymes retain a high activity at pH values one or two units away from the optimum. Most of the meat cathepsins are able to hydrolyze at least some proteins of the myofibrils. Proteolysis by cathepsins has been regarded as one of the factors responsible for tenderization of meat. However, not all of these enzymes can have a significant effect on the meat proteins in the conditions of postslaughter handling and chilling of the animal carcasses.

The muscles also contain nonlysosomal proteinases capable of hydrolyzing several myofibrillar proteins. The calcium-activated neutral proteinase (CANP) is a metalloproteinase requiring cysteinyl thiol group for activation. The highest activity of CANP against isolated myofibrils is in the pH range of 7.0–7.5, while at pH 6 and 4.5, the activity is about 80% and 40% of the maximum value, respectively. The enzyme has been isolated in a form requiring millimolar concentrations of Ca^{2+} for activation, i.e., 0.1–0.5 mM, the maximal activity being 1–2 mM Ca^{2+}. However, a second form of the enzyme has also been detected, which is activated at micromolar concentrations of calcium ions. Due to the loss of the calcium-retaining ability of the reticulum in the muscle postmortem, the concentration of the Ca^{2+} in the sarcoplasm may increase to millimolar levels.

The proteolytic enzymes in fish affect the sensory quality of various salted fish products. Furthermore, high activity of the proteases during the spawning migration may bring about changes, increasing the emulsifying ability of the meat of salmon during the spawning period (Kawai et al., 1990).

Squid muscles contain very active proteinases. This high proteolytic activity causes extensive degradation of the myofibrillar proteins in the course of fractionation. There are significant differences in the auto-proteolytic activity in the muscles of different species of squid.

Fish muscles contain proteinases responsible for the degradation of fish gels at around 60°C. These are serine proteinases of very strong activity against the myosin heavy chain. There are two types of these proteins – the sarcoplasmic enzymes and enzymes associated with the myofibrils (Toyohara et al., 1990).

Trimethylamine demethylase, present in the muscles of many gadoid fishes, affects the functional properties of fish proteins indirectly by catalyzing the formation of formaldehyde which may participate in crosslinking.

$$H_3C-\overset{\overset{\displaystyle CH_3}{|}}{\underset{\underset{\displaystyle CH_3}{|}}{N}}\rightarrow O \xrightarrow{\text{demethylase}} \overset{\overset{\displaystyle CH_3}{|}}{\underset{\underset{\displaystyle CH_3}{|}}{NH}} + HC\overset{\displaystyle O}{\underset{\displaystyle H}{\diagdown}}$$

(6.10)

These reactions are most important in frozen stored gadoid fish (Sikorski and Kołakowska, 1994).

MODIFICATION OF ISOLATED PROTEINS

Enzymatic processes may be preferred over chemical reactions for modifying functional properties of proteins to avoid undesirable side products (Matheis and Whitaker, 1987).

Calcium-activated transglutaminase catalyzes the transfer of different acyls on amino groups in proteins:

$$\text{Prot}-(CH_2)_2-CONH_2 + H_2NR \xrightarrow[Ca^{+2}]{\text{transglutaminase}} \text{Prot}-(CH_2)_2-CONHR + NH_3$$

(6.11)

This reaction can lead to "tailor-made" protein preparations, e.g., edible films of defined barrier properties from whey proteins and to covalent binding of saccharides to plant proteins rich in glutamine residues (Colas et al., 1993). Transglutaminase activity is thought to be responsible for the formation of $\epsilon(\gamma$-glutamyl)lysine cross-links in dried fish (Kumazawa et al., 1993), in frozen-stored surimi (Haard et al., 1994), and in polymerization of myosin heavy chain during setting of surimi in the manufacture of kamaboko (Kumazawa et al., 1995). Ca^{2+}-independent transglutaminase derived from *Streptoverticillium*-induced gelation of glycinin and legumin at 37°C, and the gels were more rigid and elastic than thermally induced gels (Chanyongvorakul et al., 1995).

Wheat lipoxygenase and soybean lipoxygenase, catalyzing oxidation of fatty acids, generate oxidized reaction products that improve the dough-forming properties and baking performance of the flour. A similar role is performed by polyphenol oxidase and peroxidase.

Enrichment of proteins in specific a.a.'s can be achieved in the plastein reaction, i.e., protease catalyzed transpeptidation in concentrated solutions of a.a. ethyl esters and protein hydrolyzates (Figure 6.7). Incubation of a 30–40% protein hydrolyzate with ethyl esters of, e.g., lysine, methionine, or tryptophan, with an appropriate endopeptidase at pH 4–7 at about 37°C, results after a few days in accumulation of peptides of 2–3 kDa enriched in the respective a.a. Also, plasteins free of phenylalanine residues can be obtained for phenylketonuric patients. The rate of incorporation of a.a.'s into the plastein increases with the hydrophobicity of the a.a.'s. Thus, selective removal of hydrophobic a.a. from the hydrolysate and decrease of its bitter taste is possible.

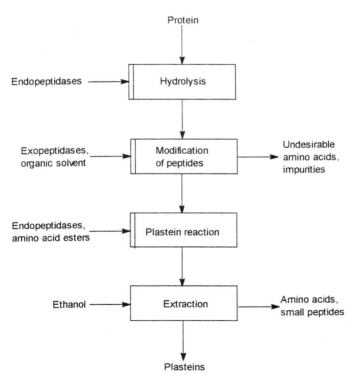

Figure 6.7 Utilization of the plastein reaction.

Cyclic adenosine monophosphate dependent protein kinase is useful for phosphorylation of a.a residues in mild conditions. The modification makes the soybean proteins soluble and media-rich in calcium and improves their emulsifying properties (Seguro and Motoki, 1990).

In a papain-catalyzed reaction, at room temperature, proteins can be acylated and enriched in SH groups using N-acetyl-homocysteinethiolactone (Sung et al., 1983):

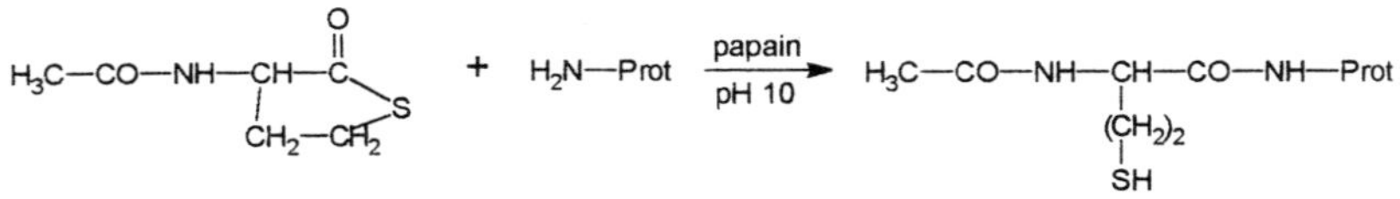

$$H_3C-CO-NH-\underset{\underset{CH_2-CH_2}{|}}{CH}-\overset{\overset{O}{||}}{C}_S + H_2N-Prot \xrightarrow[pH\ 10]{papain} H_3C-CO-NH-\underset{\underset{\underset{SH}{|}}{(CH_2)_2}}{CH}-CO-NH-Prot$$

(6.12)

CHEMICAL MODIFICATIONS

ALKYLATION

The carbonyl compounds formed due to autoxidation of lipids and in catabolic processes postmortem in muscles, or introduced with wood smoke can bring about undesirable effects by interacting with a.a. residues. On the other hand, some carbonyl compounds are used intentionally to modify the proteins, e.g., formaldehyde to harden the collagen dope in the manufacturing of sausage casings, to protect fodder meals used against deamination by the rumen microflora, and to immobilize enzymes on supports.

In reducing conditions, alkylation of amino, indole, thiol, and thioether groups, and binding of saccharides to a.a. residues are possible, in order to modify the functional properties of proteins intended to be used in small quantities in various food formulations:

$$2Prot-NH_2 + 4HCHO + NaCNBH_3 \xrightarrow[0^\circ C]{pH\ 9} 2Prot-N(CH_3)_2 + NaHCNBO_3 + H_2O$$

(6.13)

For alkylation of amino, phenol, imidazole, thiol, and thioether groups in a.a. of proteins, reactions with haloacetates and haloamides can be used:

$$Prot-SH + ICH_2COOH \longrightarrow Prot-S-CH_2COOH + IH$$

(6.14)

$$Prot—NH_2 + ICH_2CONH_2 \longrightarrow Prot—NH—CH_2—CONH_2 + IH$$

(6.15)

The digestibility of such derivatives is generally somewhat lower than that of the unmodified proteins.

Malonaldehyde, a typical lipid oxidation product, can form cross-links in proteins by reacting with two amino groups:

$$OHC—CH_2—CHO + 2Prot—NH_2 \longrightarrow Prot—N=CH—CH_2—CH=N—Prot + H_2O$$

(6.16)

There is also a possibility that aliphatic monoaldehydes generated in lipid autoxidation can participate in cross-linking of proteins. Hexanal forms cross-links between two N^α-acetyltryptophan molecules at 50°C by reacting at the 1- and 2- position of the indole ring (Kaneko et al., 1989):

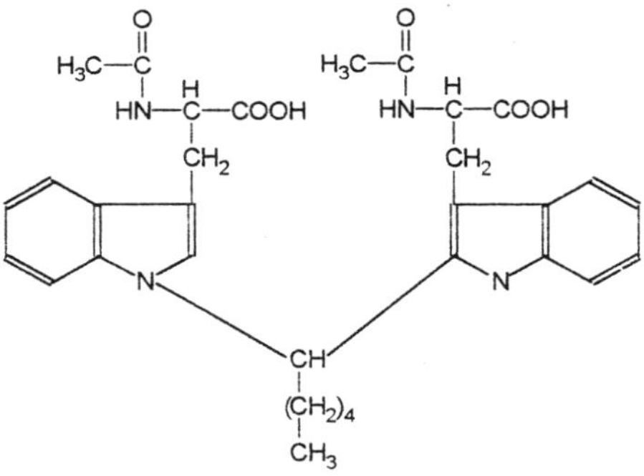

(6.4)

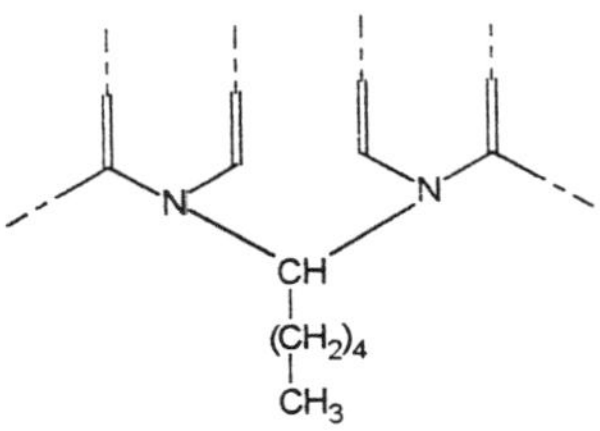

(6.5)

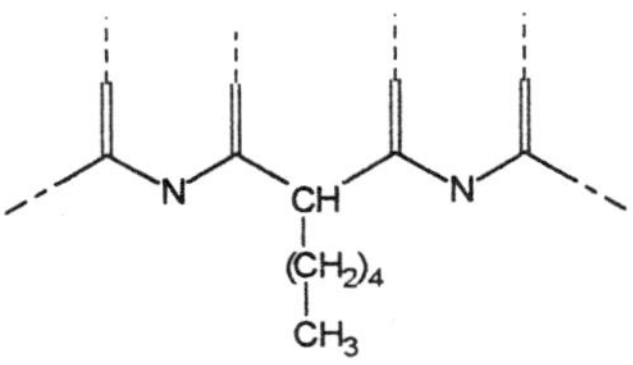

(6.6)

Formaldehyde can react in food systems with amino, amide, hydroxyl, and thiol groups in a.a. residues of proteins, even at room temperature. Some of these reactions also result in cross-linking of the proteins. In cheese, the biogenic formaldehyde reacts with the N-terminal histidine of γ_2-casein to form spinacine:

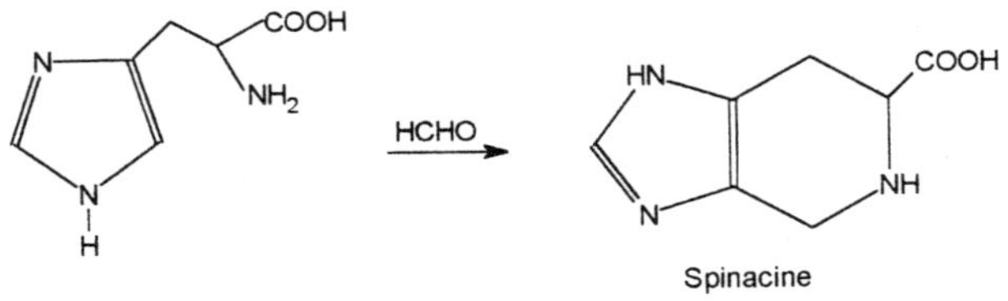

Spinacine

(6.17)

Other carbonyl compounds produced in cheese by the lactic acid bacteria form other derivatives in reactions with histidine (Pellegrino and Resmini, 1996).

ACYLATION

Acylation of nucleophilic groups is intended to increase the hydrophobicity of the protein, introduce additional ionizable carboxyl groups or a.a. residues, or contribute to cross-linking:

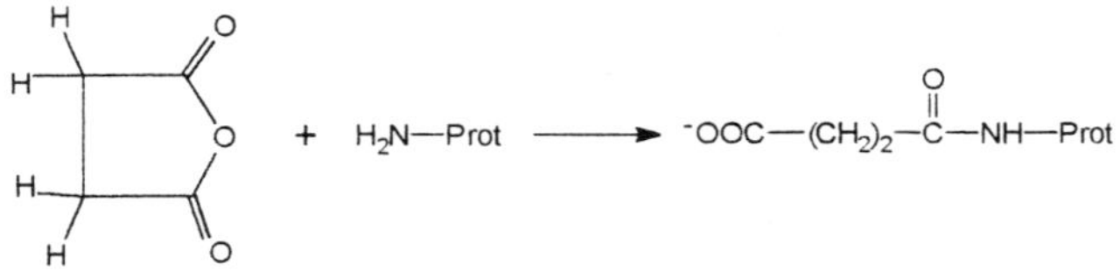

(6.18)

The acylating agents may react with amino, imidazole, hydroxyl, phenol, and thiol groups in protein a.a. residues. The rate of reaction depends upon the properties of the nucleophilic groups, pH, the steric factors resulting from the protein conformation, and the presence of inhibitors. The amino and tyrosyl groups can be acylated easily; histidine and cysteine derivatives hydrolyze readily, while hydroxyl groups do not react easily in aqueous solutions. Acylation of hydroxyl groups in a.a. residues of proteins proceeds in the presence of an excess of acetic anhydride only after the amino groups have been acylated. For the formation of isopeptide linkages, N-carboxyanhydrides of a.a.'s are suitable.

Some food proteins are rich in phosphoric acid residues. The acid may either form ester bonds with serine residues, as in caseins and in egg proteins, or may stabilize the native conformation of protein micelles by electrostatic interactions with negatively charged groups and calcium ions, as in the caseins. In soy proteins, serine and threonine residues can be esterified and lysine amidated with cyclic sodium trimetaphosphate at pH 11.5 and 35°C (Sung et al., 1983):

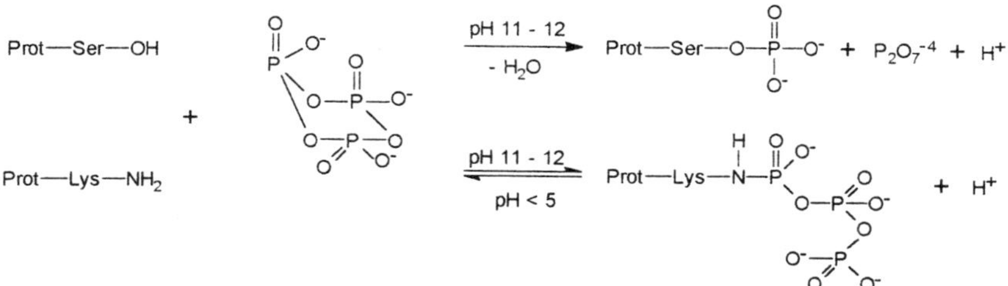

(6.19)

The N^ϵ-triphospholysine residues protect the NH_2 group during heating in an alkaline environment and hydrolyze at pH < 5. For chemical phosphorylation, other reagents may be used, e.g., phosphorus oxychloride, phosphorus pentoxide, etc. The reactivity of these compounds may, however, lead to undesirable cross-linking of the proteins:

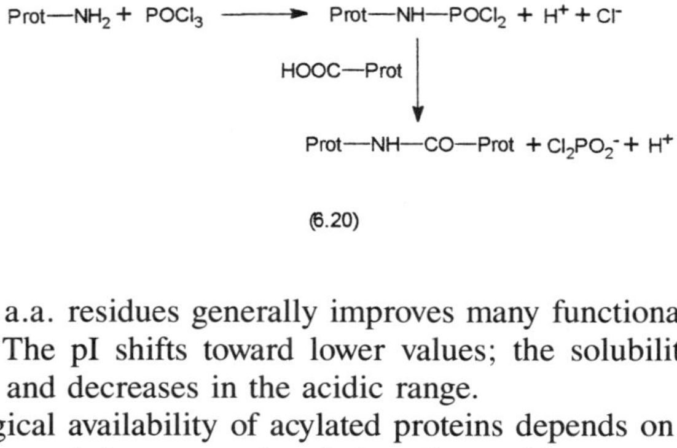

(6.20)

Acylation of a.a. residues generally improves many functional properties of proteins. The pI shifts toward lower values; the solubility increases above the pI and decreases in the acidic range.

The biological availability of acylated proteins depends on the kind of the acyl group and the extent of acylation. Isopeptides are generally well utilized, while the availability of other derivatives decreases with the size of the acylating moiety.

N-NITROSATION

The nitrous acid generated in foods at low pH from endogenous or added nitrites decomposes easily to yield the nitrosating agents nitrous anhydrate and nitrosonium ion:

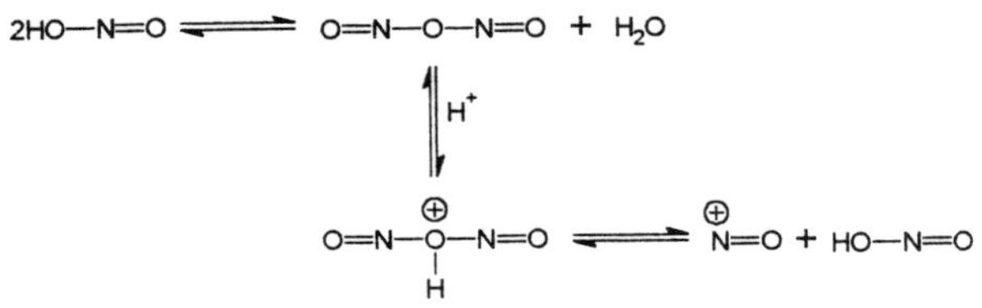

$$2HO\!-\!N\!=\!O \rightleftharpoons O\!=\!N\!-\!O\!-\!N\!=\!O + H_2O$$

$$\Big\updownarrow H^+$$

$$O\!=\!N\!-\!\underset{H}{\overset{\oplus}{O}}\!-\!N\!=\!O \rightleftharpoons \overset{\oplus}{N}\!=\!O + HO\!-\!N\!=\!O$$

(6.21)

Reactions of these compounds with the secondary and tertiary amines contained in many foods lead to the known carcinogens N-nitrosoamines:

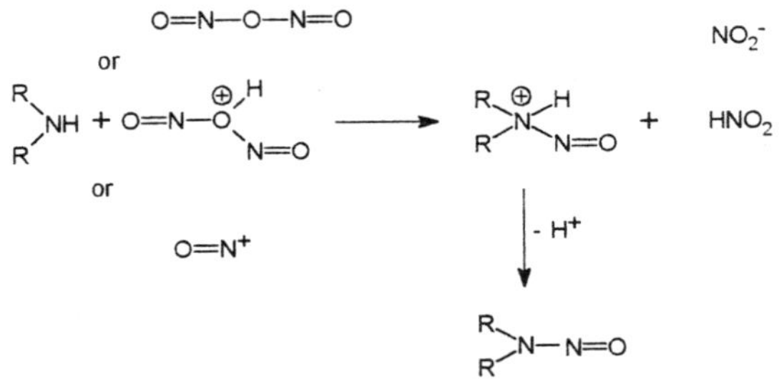

(6.22)

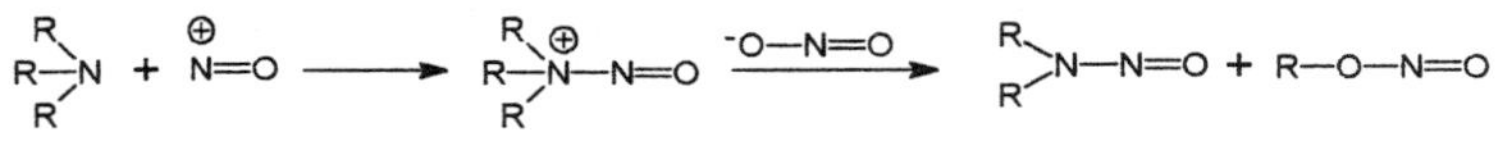

(6.23)

The rate of N-nitrosation increases with the pK_a of the amine and depends on the pH — it is highest at pH 2–4. The reaction can be inhibited by compounds capable of binding the nitrosating agents — in meat curing, sodium ascorbate is very effective. Foods low in amines and nitrites contain generally about 1–10 ppb, while cured and heavily smoked meat and fish up to several hundred ppb of N-nitroso compounds.

REACTIONS WITH PHOSPHATES

Proteins may, at acid pH, form protein–phosphate complexes of low solubility. The metaphosphates, e.g., $NaPO_3$, which are less hydrated than the ortophosphates, e.g., Na_3PO_4, form complexes that are less soluble than complexes with ortophosphates. This has been utilized for the modification of functional properties of protein concentrates and preparations, for the separation of proteins in different food processing operations, and in treatment of protein-containing food plant effluents.

Polyphosphates improve the sensory quality of many food products, e.g., by preventing the separation of the butter fat and aqueous phases in evaporated milk or of the formation of gel in concentrated milk sterilized by HTST, by stabilizing the fat emulsion in processed cheese by disrupting the casein micelles and thus enhancing the hydrophobic interactions between lipids and casein submicelles, and by increasing the WHC and improving the texture of many cooked meat products. The mechanisms involved in different applications depend upon the properties of the phosphates and of the products, as well as on the parameters of processing.

In meat products, the increase in WHC, texture improvement, and decrease in drip may be caused by increasing the pH, complexing of Ca^{2+} and Mg^{2+}, binding of phosphates to proteins, rupturing the cross-links between myosin and actin, and dissolving part of the proteins (Hamm, 1970). Phosphates facilitate the manufacture of meat products of low sodium content. To produce an emulsified meat product of acceptable quality, at least 2.5% salt is required. The contents of salt can be decreased to 1.5–2.0% without loss in product quality, by adding phosphates in amounts of 0.35% to 0.5%. Generally, proprietary blends of several polyphosphates are used, to give best quality and prevent precipitation of orto- and pyrophosphates in brines rich in Ca^{2+}. The polyphosphates that are most effective in stabilizing the meat batters also have the most retarding effect on the formation of the desirable meat color. The decrease in the rate of color formation is probably partially caused by the pH effect of the polyphosphates (Knipe, et al., 1988).

Polyphosphates are used as additives in frozen fish minces. They are also applied in form of dips to fish fillets prior to freezing to prevent thaw drip losses, and to improve the texture of canned fish. Mainly about 10% solutions of $Me_5P_3O_{10}$ and $Me_4P_2O_7$ are used for 1–2 min. Also, different proprietary mixtures are applied, e.g., $Na_4P_2O_7 + Na_2H_2P_2O_7$ or $Na_3PO_4 + Na_4P_2O_7 + Na_2H_2P_2O_7$.

The widespread use of phosphate food additives have caused, in recent years, an increase in the contents of phosphorus in many food products. An excess of phosphorus in the diet may disturb the optimum Ca:P ratio.

USE OF PROTEINS AS FUNCTIONAL COMPONENTS IN FOODS

LEGUME PROTEINS

Protein isolates are used in foods because of their functional properties. A large range of applications has been found mainly for protein isolates from soybeans and other legumes, for different milk proteins, and for egg albumen.

Soybean proteins are used as a variety of traditional products, i.e., soymilk or fermented items, and as defatted flour, grits, concentrates, or isolates. The traditional products are prepared according to procedures known in the Orient, involving water extraction, cooking, coagulation of the proteinaceous curd, roasting, and fermentation, often in combinations with other components. The different forms of concentrated or isolated protein products are prepared by milling, toasting, extraction of fat and saccharides, and isolation of protein fractions. The antinutritional factors present in soybeans are usually inactivated or removed in processing. The "beany" or "painty" off-flavor in soymilk due to volatiles generated in lipoxygenase catalyzed reactions is prevented by thermal denaturation of the enzyme before or during grinding with water (Kwok and Niranjan, 1995). The protein isolates may be tailor-made to have the desirable functional properties.

Grits, flours, and isolates for food applications are also produced from other legumes, mainly peanuts, beans, broad beans, and peas.

MILK PROTEINS

Because of a very high biological value and the absence of antinutritional factors, except for some allergenic activity, milk proteins have found various applications in formulated foods, also as meat extenders. Initially, only the caseins were used, but recently the recovery of whey proteins and their fractions is economically feasible. Least soluble is acid casein. Its solubility and foaming properties can be improved by glycosylation of the ϵ-NH_2 groups (Closs et al., 1990). Sodium and calcium caseinates obtained by acid precipitation and neutralization are soluble, very heat stable, and have high WHC, emulsifying, foaming, and gelling properties. Rennet casein, richer in calcium, has low solubility in the presence of Ca^{2+}. Precipitates produced by heat denaturation of the whey proteins and coprecipitation with casein by addition of acid or calcium salts are more soluble than the acid and rennet casein and have higher nutritional value. Different whey protein concentrates and isolates are produced by complexing with phosphates or polysaccharides, by gel filtration, ultrafiltration, or heat denaturation. At a pH of about 4.2 and at 55–65°C, α-lactalbumin undergoes isoelectric precipitation due to dissociation of the Ca^{2+} and hydrophobic interactions (Bramaud et al., 1995). Also, other minor whey proteins precipitate while the soluble β-lactoglobulin is separated, concentrated by ultrafiltration, neutralized, and spray-dried. The product has superior water-holding and gelling properties in meat products. Other undenatured forms of whey proteins are also used as foam stabilizers and gelling agents, and some are soluble under acid conditions. The purified α-lactalbumin fraction is more suitable for infant food formulations than whey protein concentrate since human milk does not contain β-lactoglobulin.

EGG PROTEINS

Egg albumen is used in various food formulations because of its foaming properties and heat-gelling ability, while egg yolk serves as an emulsifying agent. The gel strength of egg white proteins can be increased by preheating the dry protein at 80°C before use (Kato et al., 1990a). In the gels formed by preheated egg white, the molecular weight of the aggregates is much smaller than in gels made of nonpreheated proteins. Preheating of dry egg white also brings about a decrease in the enthalpy and temperature of denaturation of the proteins and, as a result, increased flexibility of the molecules, leading to more cohesive interfacial films, i.e., improved surface functional properties. The more cohesive films composed of overlapping polypeptides are more capable of expanding under stress than the interfacial films formed from nonpreheated proteins (Kato et al., 1990b).

Egg white and egg yolk gels can be prepared without heating by applying high pressure (Hayashi et al., 1989). The albumen and the yolk from fresh eggs form stiff gels after being exposed 30 min to a pressure above 6,000 kg/cm² and 4,000 kg/cm², respectively. The pressure-induced gels are more adhesive, more elastic, and more digestible than the boiled egg.

EDIBLE COATINGS

Collagen, gelatin, casein, total milk proteins, wheat gluten, corn zein, and soy protein are used for producing edible protein films or protein-containing composite coatings applicable for different food products. The barrier properties, appearance, and tensile strength of these products depend on the characteristics of the raw materials, contents of plasticizers, and conditions of fabrication. The films are usually made by preparing a protein solution at pH values far from the pI, controlled heating, casting, and drying. The edible film prepared from commercial soy protein isolate treated with alkali at pH 8 has good barrier and mechanical properties (Brandenburg et al., 1993).

REFERENCES

Boatright, W. L. and Hettiarachchy, N. S. 1995. "Soy protein isolate solubility and surface hydrophobicity as affected by antioxidants," *J. Food Sci.*, 60:798–800.

Bramaud, C., Mimar, P. and Daufin, G. 1995. "Thermal isoelectric precipitation of α-lactalbumin from a whey protein concentrate: influence of whey-calcium complexation," *Biotechnology and Bioengineering,* 47:121–130.

Brandenburg, A. H., Weller, C. L. and Testin, R. F. 1993. "Edible films and coatings from soy proteins," *J. Food Sci.,* 58:1086–1089.

Chanyongvorakul, Y., Matsumura, Y., Monaka, M., Motoki, M. and Mori, T. 1995. "Physical properties of soy bean and broad bean 11S globulin gels formed by transglutaminase reaction," *J. Food Sci.,* 60:483–488.

Closs, B., Courthaudon, J.-L. and Lorient, D. 1990. "Effect of chemical glycosylation on the surface properties of the soluble fraction of casein," *J. Food Sci.,* 55:437–439, 461.

Colas, B., Caer, D. and Fournier, E. 1993. "Transglutaminase catalyzed glycosylation in vegetable protein; effect on solubility of pea legumin and wheat gliadins," *J. Agric. Food Chem.,* 41:1811–1815.

Damodaran, S. 1989. "Interrelationship of molecular and functional properties of food proteins," in *Food Proteins,* John E. Kinsella and William G. Soucie, eds. Champaign, IL: The American Oil Chemists' Society, pp. 21–51.

Doi, E., Shimizu, A., Oe, H. and Kitabatake, N. 1991. "Melting of heat-induced ovalbumin gel by pressure," *Food Hydrocolloids,* 5:409–425.

Dunajski, E. 1977. Abstracts of papers of the VIII Scientific Session of the Committee of Food Technology and Chemistry of the Polish Academy of Sciences. 22 pp.

Goodenough, P. W. 1995. "A review of protein engineering for the food industry," *International Journal of Food Science and Technology,* 30:119–139.

Haard, N. F., Simpson, B. K. and Pan, B. S. 1994. "Sarcoplasmic proteins and other nitrogenous compounds," in *Seafood Proteins,* Zdzisław E. Sikorski, Bonnie S. Pan, and Fereidoon Shahidi, eds. New York: Chapman & Hall, pp. 13–39.

Hamm, R. 1960. "Biochemistry of meat hydration," in *Advances in Food Research, Vol. 10,* C. O. Chichester, E. M. Mrak, and G. F. Stewart, eds. New York: Academic Press, pp. 350–463.

Hamm, R. 1970. "Interactions between phosphates and meat proteins," in *Symposium: Phosphate in Food Processing,* Chap. 5, DeMan and Melnychyn, eds. Westport, CT: AVI.

Hamm, R. 1986. "Functional properties of the myofibrillar system and their measurements," in *Muscle as Food,* P. J. Bechtel, ed. New York: Academic Press, pp. 135–199.

Hayashi, R., Kawamura, Y., Nakasa, T. and Ohinaka, O. 1989. "Application of high pressure to food processing: pressurization of egg white and yolk, and properties of gels formed," *Agric. Biol. Chem.,* 53:2935–2939.

Hoseney, R. C. and Rogers, D. E. 1990. "The formation and properties of wheat flour doughs," *CRC Crit. Rev. Food Sci. Nutr.,* 29:73–93.

Jost, R. 1993. "Functional characteristics of dairy proteins," *Trends in Food Science & Technology,* 4:283–288.

Kaneko, S., Okitani, A., Hayase, F. and Kato, H. 1989. "Novel cross-linking compounds formed through the reaction of N^α-acetyltryptophan with hexanal," *Agric. Biol. Chem.,* 53:2679–2685.

Kantha, S. S., Watabe, S. and Hashimoto, K. 1990. "Comparative biochemistry of paramyosin—A review," *J. Food Biochem.,* 14:61–88.

Kato, A., Ibrahim, H. R., Takagi, T. and Kobayashi, K. 1990a. "Excellent gelation of egg white preheated in the dry state is due to a decreasing degree of aggregation," *J. Agric. Food Chem.*, 38:1868–1872.

Kato, A., Ibrahim, H. R., Watanabe, H., Houma, K. and Koboyashi, K. 1990b. "Enthalpy of denaturation and surface functional properties of heated egg white proteins in the dry state," *J. Food Sci.*, 55:1280–1283.

Kawai, Y., Hirayama, H. and Hatano, M. 1990. "Emulsifying ability and physiochemical properties of muscle proteins of fall chum salmon *Oncorhynchus keta* during spawning migration," *Nippon Suisan Gakkaishi*, 56:625.

Killday, K. B., Tempesta, M. S., Bailey, M. E. and Metral, C. J. 1988. "Structural characterization of nitrosylhemochromogen of cooked cured meat: Implications in the meat-curing reaction," *J. Agric. Food Chem.*, 36:909–914.

Knipe, C. L., Olson, D. G. and Rust, R. E. 1988. "Effects of inorganic phosphates and sodium hydroxide on the cooked cured color, pH, and emulsion stability of reduced-sodium and conventional meat emulsions," *J. Food Sci.*, 53:1305.

Kumazawa, Y., Seguro, K., Takamura, M. and Motoki, M. 1993. "Formation of ε(gamma-glutamyl)lysine cross-link in cured horse mackerel meat induced by drying," *J. Food Sci.*, 58:1086–1089.

Kumazawa, Y., Numazawa, T., Seguro, K. and Motoki, M. 1995. "Suppression of surimi gel setting by transglutaminase inhibitors," *J. Food Sci.*, 60:715–771.

Kwok, K. C. and Niranjan, K. 1995. "Review: Effect of thermal processing on soymilk," *International Journal of Food Science and Technology*, 30:263–295.

Lan, Y. H., Novakofski, J., McCusker, R. H., Brewer, M. S., Carr, T. R. and McKeith, F. K. 1995. "Thermal gelation properties of protein fractions from pork and chicken breast muscles," *J. Food Sci.*, 60:742–747.

Mannherz, H. G. 1992. "Crystallization of actin in complex with actin-binding proteins," *J. Biol. Chem.*, 267(17):11661–11664.

Matheis, G. and Whitaker, J. R. 1987. "A review: Enzymatic cross-linking of proteins applicable to foods," *J. Food Biochemistry*, 11:309–327.

Mulvihill, D. M. and Kinsella, J. E. 1987. "Gelation characteristics of whey proteins and β-lactoglobulin," *Food Technology*, 41:102–111.

Nakai, S. and Li-Chan, S. 1988. *Hydrophobic Interactions in Food Systems*, Boca Raton: CRC Press.

Nishimura, K., Ohishi, N., Tanaka, Y., Ohgita, M., Takeuchi, Y., Watanabe, H., Gejima, A. and Samejima, E. 1992. "Effects of ascorbic acid on the formation process for a heat-induced gel of fish meat (kamaboko)," *Biosci. Biotech. Biochem.*, 56:1737–1743.

Oakenfull, D. 1987. "Gelling agents," *CRC Crit. Rev. Food Sci. Nutr.*, 26:12.

Pellegrino, L. and Resmini, P. 1996. "Evaluation of the stable reaction products of histidine with formaldehyde or with other carbonyl compounds in dairy products," *Z. Lebensm. Unters. Forsch.*, 202:66–71.

Price, J. L., Lyons, C. E. and Huang, R. C. C. 1990. "Seasonal cycle and regulation of temperature of antifreeze protein mRNA in a Long Island population of winter flounder," *Fish Physiology and Biochemistry*, 8:187–198.

Regenstein, J. M. and Regenstein, C. E. 1984. *Food Protein Chemistry*, Orlando: Academic Press, Inc.

Seguro, K. and Motoki, M. 1990. "Functional properties of enzymatically phosphorylated soybean proteins," *Agric. Biol. Chem.*, 54:1271–1274.

Shinizu, A., Kitabatake, N., Higasa, T. and Doi, E. 1991. "Melting of the ovalbumin gels by heating: Reversibility between gel and sol," *Nippon Shokuhin Kogyo Gakkaishi,* 38:1050–1056.

Sikorski, Z. E. and Kołakowska, A. 1994. "Changes in proteins in frozen stored fish," in *Seafood Proteins,* Zdzisław E. Sikorski, Bonnie S. Pan, and Fereidoon Shahidi, eds. New York and London: Chapman & Hall, pp. 99–112.

Sung, H. Y., Chen, H. J., Liu, T. Y., and Su, J. Ch. 1983. "Improvement of the functionalities of soy protein isolate through chemical phosphorylation," *J. Food Sci.,* 48:716–721.

Tani, F., Murata, M., Higasa, T., Goto, M., Kitrabatake, N. and Doi, E. 1993. "Heat-induced transparent gel from hen egg lysozyme by a two-step heating method," *Biosci. Biotech. Biochem.,* 57:209–214.

Toyohara, H., Kinoshita, M. and Shimizu, Y. 1990. "Proteolytic degradation of threadfin-bream meat gel," *J. Food Sci.,* 55:259–260.

Trziszka, T. 1994. "The functional properties of egg white," in *Egg and Egg Products* (in German), W. Ternes, L. Acker, and S. Scholtyssek, eds. Berlin: Paul Parey, pp. 238–255.

Wang, Ch. H. and Damodaran, S. 1990. "Thermal gelation of globular proteins: weight-average molecular weight dependence of gel strength," *J. Agric. Food Chem.,* 38:1157–1164.

Whitaker, J. R. 1977. "Enzymatic modification of proteins applicable to foods," in *Food Proteins. Improvements through Chemical and Enzymatic Modification,* R. E. Feeney and John R. Whitaker, eds. Advances in Chemistry Series 160, Washington D.C.: American Chemical Society, p. 97.

Zehetner, G., Bareuther, Ch., Henle, T. and Klostermeyer, H. 1995. "Inactivation kinetics of gamma-glutamyltransferase during the heating of milk," *Z. Lebensm. Unters. Forsch.,* 201:336–338.

Zhu, H. and Damodaran, S. 1994. "Effects of calcium and magnesium ions on aggregation of whey protein isolate and its effect on foaming properties," *J. Agric. Food Chem.,* 42:856–862.

CHAPTER 7

Rheological Properties of Food Systems

TADEUSZ MATUSZEK

INTRODUCTION

THE food functionality in general can be expressed through the hurdle technological effect, which was achieved at the final stage of food processing. Each particular food product can be assessed by a certain set of hurdles that differs in quality and intensity, depending on the technology used (Gorris, 1995). The hurdles might influence the stability and the sensory, nutritive, technological, and economic properties of a product.

The hurdles affecting the shelf life of foods also influence other food properties, including the texture. The effects of several physical, chemical, and mechanical treatments should be carefully considered in developing new processes and products. It is not enough to describe the composition of a food product and to determine the conditions and types of unit operation necessary to achieve the required quality. It should be examined how the major food ingredients such as water, salt, hydrocolloids, starches, lipids, proteins, flavors and additives interact with each other and affect the product quality with respect to the microstructure, texture, and appearance.

From the food engineering point of view, food functionality is the specific response of foods to applied forces encountered during the preparation, processing, storage, and consumption of food (Kokini et al., 1993). The understanding of food at the molecular level involves the application of both theoretical and experimental techniques of chemistry, physics, mathematics, fluid mechanics, biochemistry, and biophysics to understand how the molecular properties and interactions affect the final quality of the product. If the texture is to be controlled, then the effect of individual components of the formulations should be known.

The forces affecting foods during engineering operations may include various forms of energy applied. It can be, for instance, transport fluxes related to the mass and heat flow; electromagnetic energy including light, microwave, and infrared radiation; or chemical reactions, which are responsible for transferring the food from one thermodynamic state to another through changes in enthalpy, entropy, and resulting free energies.

The energy deposited and the resulting forces usually expose their effects at any level in the hierarchy of food structure, i.e., from the molecular level to the formation of phases, networks, aggregates, cells, and finally, to the food products themselves.

Many food systems are formed in conditions far away from thermodynamic equilibrium because all food components can potentially interact chemically with one another to varying degrees. Most food materials and their complex systems are biologically active and physically unstable with continuous changes of their structure. Those physical, chemical, and mechanical instabilities in many food systems are a direct result of the nonequilibrium nature of these systems. During food processing, the raw material has its chemical and physical properties extensively altered. This results in a final product, whose appearance is very different from the original, primary native microstructure. Most processing operations are strongly related to the structural aspects of water as a main solvent and plasticizer and to the contribution of water to hydrodynamic properties of the food systems. The thermodynamic aspects describing water relation in equilibrium with its surroundings at certain relative humidity, pressure, and temperature should also be considered. Many systems involve complex mixtures of molecules in low-water environments, and interfacial phenomena with all forces of intermolecular interactions play a very important role in food functionality. Therefore, the kinetic and thermodynamic aspects of possible physical and chemical interactions have to be investigated to obtain a complete understanding of all dynamic processes leading to food functionality (Figure 7.1).

THE FLOW PROPERTIES OF LIQUID AND SEMI-SOLID FOOD SYSTEMS

The flow properties are critical for attaining accuracy and quality in designing food processing machines and food industry plants. They are also vital in modeling the processing operations. Various energy and heat treatments of food include thermal conductivity, thermal diffusivity, density, consistency and concentration changes, specific enthalpy, specific heat, texture, and mechanical and rheological properties.

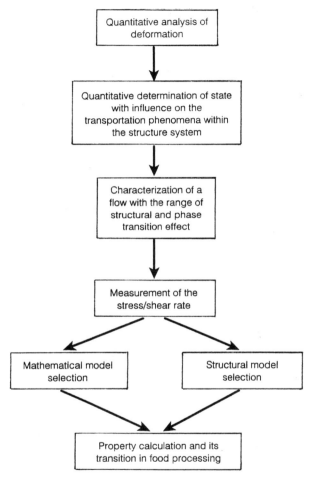

Figure 7.1 An outline of general methodology of the functional food properties seen as deformation and property changes (Matuszek, 1995).

Foods are mostly nonhomogeneous, and structure formation means that various states of aggregation occur during processing. The engineering phase of processing usually involves changes in the structure of the raw materials. The food technologist and engineer would like to be able to predict the properties of any formulation of components over a wide range of processing and storage conditions. In order to make this happen, it is necessary to understand more about factors that affect

- all levels of food structure organization
- the kinetics of competitive adsorption of the food components at fluid interfaces
- the developing time-dependent properties of food formulations
- fouling problems and their effects on processing performance for different types of membranes that are created within the microstructure of the systems

All these factors are related to the theory of flow for predicting the rates of molecular transport and their relationships to molecular and consequently to food microstructure properties. This molecular flow information leads to a better understanding of surface rheological factors and to the organization and motions together with intermolecular interaction of molecules at interfaces. Intermolecular forces are responsible for many of the bulk properties of foods. A realistic processing description of the relationship among pressure, volume, temperature, energy, and other properties of the material must include the effects of attractive and repulsive forces between molecules. Repulsive forces prevent the molecules from approaching one another too closely and account for the low compressibility of liquids. Intermolecular forces between near and distant neighbors dictate the ordered molecular arrangements in crystalline solids. These forces also account for the elasticity of solids and for a very complex, condensed phase. Furthermore, the properties of structure influenced by velocity of propagation of disturbances in it, such as local density, temperature changes, and kinetic and temperature instabilities, are neither constant nor uniform. The time and energy variation for all occurrences in the various regions of food structure correspond to their conversion to flow properties like viscosity, surface tension, and diffusion of liquids through membranes and other barriers.

In food products that are complex physiochemical systems containing various types of solutions, as well as fibrous, cellular, and crystalline components, the relationship shown in Figure 7.2 exists. It follows that the study of food texture involves several areas:

- the structure, in terms of both micro- and macrostructure
- the evaluation of the rheological properties considered as physical properties

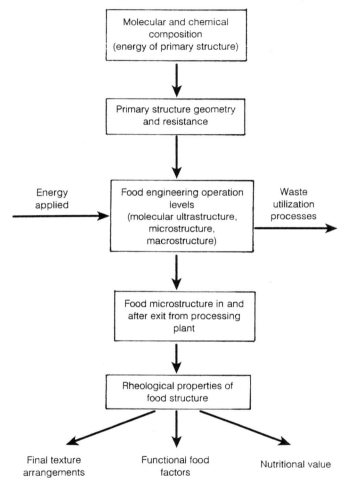

Figure 7.2 Evolution of food structure arrangements from raw materials and a process used to transform them into products.

- the evaluation of rheological and textural properties by the human sense organs
- the interrelationship of the physical and sensory measurements of food texture

In this respect, the texture comprises all physical characteristics of foods related to response to applied force and measured objectively in terms of force, distance, and time. Texture depends on the various constituents and structural elements of foods in which the microstructure components are formed and then clearly recognized in terms of flow and deformation during different processing treatment.

In general, food can be classified as Newtonian and non-Newtonian. The similarity for Newtonian liquids is represented by a straight line through the origin. It means that the measurements of the viscosity coefficient necessitate only the measurements of one point on this line. For Newtonian viscosity depends only on the temperature and pressure, and only a few foods fall into this category. Newtonian liquid food products include aqueous solution of sucrose, dextrose and low-molecular-weight saccharides, invert sugar syrups, molasses, carbonated soft drinks, alcoholic beverages, and other aqueous solutions of low-molecular-weight substances. Vegetable oils and cooking fats in the liquid form, milk, and syrups are generally Newtonian. The viscosity of these solutions is related to the concentration of the dissolved phase and to the temperature.

Most fluid and semi-solid foods of practical interest are non-Newtonian. Their viscosity depends strongly on the shear rate (Figure 7.3). Shear stress versus shear rate curves may have the shape represented by curves numbered from 1 to 7 (Figure 7.4).

Some of the most difficult material properties of fluid and semi-solid foods to determine experimentally are viscometric functions or steady shear rheological properties. The flow properties of a liquid and semi-solid food system should be measured in the following instances:

- when new food products have to be manufactured with clearly specified rheological characteristics
- when rheological properties must be carried through various steps of engineering operations without any changes
- when rheological properties indicate directly or indirectly the quality of the final product
- when rheological properties chosen to describe the texture of the product appear to be the most successful on the market
- when continuous process evaluation of the rheological properties plays an essential role in development of new products and designs better equipment

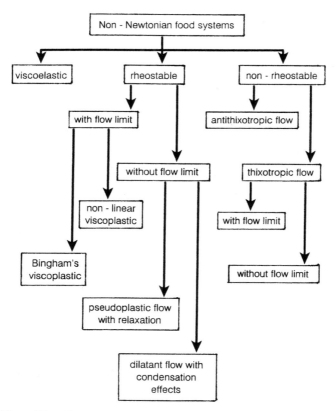

Figure 7.3 Different types of non-Newtonian liquids and semi-solid foods.

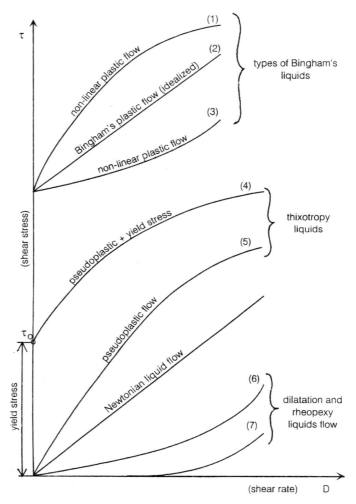

Figure 7.4 Relationship between shear stress and shear rate for different food's flow characteristics.

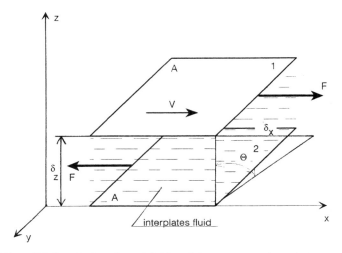

Figure 7.5 Sample shear flow: definition of shear stress, strain, and shear rate.

There are three main rheological properties of materials, i.e., viscous flow, plastic flow, and elastic deformation. Elastic materials also have a stress deformation behavior represented by a straight line through the origin. However, in this case, the deformation is reversible upon removal of the force. Many foods show time-dependent rheological behavior and a combination of these, such as viscoelasticity and viscoplasticity.

To characterize Newtonian and non-Newtonian food properties, several approaches can be used, and the whole stress-strain curve can be obtained. One of the most important textural and rheological properties of foods is viscosity (or consistency). The evaluation of viscosity can also be demonstrated by reference to the evaluation of creaminess, spreadability, and pourability characteristics. All of them depend largely on shear rate and are affected by viscosity.

If the thin layer of liquid between two small planes of each of area A (Figure 7.5) is considered to be part of a laminar flow, it has a greater velocity and pulls the middle layer forward with an equal force *F,* while the layer below, which has lesser velocity, pulls the middle layer back with an equal force *F.* These two equal, opposite, and parallel forces form a shear couple and produce a shear stress of magnitude *F/A*. Newton suggested that the shear stress was directly proportional to the velocity gradient, i.e,

$$F/A \propto dV/dz \qquad (7.1)$$

$$\therefore F/A = \eta dV/dz \qquad (7.2)$$

or

$$\tau = \eta \cdot D \tag{7.3}$$

where η is the constant of proportionality and is called the dynamic viscosity and $dV/dz = D$ is called the velocity gradient, or shear rate. The two planes are separated by a distance ∂z, and the shear strain $\partial \theta$ is produced in time ∂t. If ∂x is not large, then $\partial \theta = \partial x / \partial z$. Because this has taken place in time ∂t, therefore, the rate of change of shear strain is $\partial \theta / \partial t \cong dV/dz = $ velocity gradient, where dV is the velocity of the upper layer relative to the lower.

Newtonian liquids show a stress-strain relationship represented by a straight line (Figure 7.6). In Figure 7.6(b) there are viscosities of two Newtonian fluids at different shear rates. These two materials represented by 1 and 2 may have the same apparent viscosity when there will be no direct proportionality between shear stress and rate of shear. The flow behavior for non-Newtonian stress-strain relationship is represented by different curves (Figure 7.7). The term non-Newtonian is applied to all materials that do not obey the direct proportionality between shear stress and rate of shear. For a non-Newtonian fluid, the viscosity has no meaning, unless the shear rate is specified, and the apparent viscosity is not constant. Apparent viscosity (η_{app}) can be used for easy comparison between Newtonian and non-Newtonian fluids at particular shear rate. It is usually defined as the ratio of the shear stress over the rate of shear.

During food engineering operations, many fluids deviate from laminar flow when subjected to high shear rates. The resulting turbulent flow gives

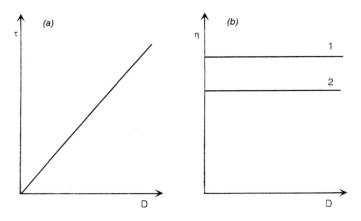

Figure 7.6 (a) Newtonian liquid; (b) viscosities of two Newtonian (1,2,) fluids at different shear rates.

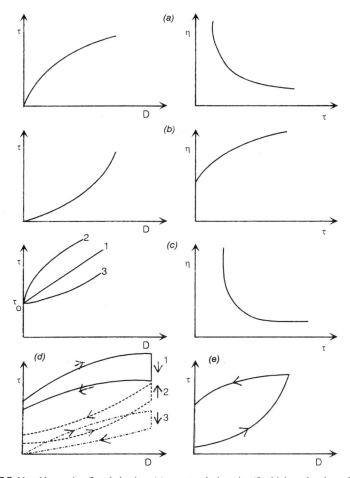

Figure 7.7 Non-Newtonian flow behavior: (a) structural viscosity (for high molecular solution); (b) dilatant flow (suspension with high concentration); (c) viscoplastic with flow limits: 1—ideal plastic, 2,3,—non-linear plastic flow, (d) thixotropy flow (1), antithixotropy (2), and viscoelastic flow (3); (e) rheopexy flow.

rise to an apparent increase in viscosity as the shear rate increases in laminar flow, i.e., shear stress = viscosity × shear rate. In turbulent flow, it would appear that: total shear stress = (laminar stress + turbulent stress) × shear rate. The most important part of turbulent stress is related to the eddies' diffusivity of momentum. This can be recognized at the atomic-scale mechanism of energy conversion and its redistribution to the dynamics of mass transport processes, responsible for the spatial and temporal evolution of the food system.

In general, there are three main types of non-Newtonian liquids and semi-solids:

- time-independent—for which the rate of shear depends only on the shear stress
- time-dependent—for which the relationship between the rate of shear and the shear stress depends on the time of shear
- viscoelastic—which has characteristics of both elastic solids and viscous liquids

In time-independent liquid food products, the flow curve is linear but intersects the shear stress axis at a positive value of shear stress. This value is known as a yield stress. The significance of the yield stress is that it is the stress that must be exceeded before the material will flow. This type of flow can be characterized by the rheological equation (for the Bingham-Schwedoff model)

$$\tau = \eta_{pl} \cdot D + \tau_0 \tag{7.4}$$

where

τ_o = yield stress
η_{pl} = Bingham plastic viscosity

To the same family of curves belong pseudoplastic materials. This fluid shows a decrease in apparent viscosity with an increase in rate of shear and is typical of the majority of non-Newtonian liquid food products. The way most often used to describe the properties of these materials is an empirical Ostwald-de Waele power-law equation

$$\tau = k \cdot D^s \tag{7.5}$$

where

k = consistency index
s = flow behavior index, which for pseudoplastic materials is less than 1 and greater than zero ($0 < s < 1$).

This equation can be used to describe the rheological properties of any time-independent fluid if applied over a limited range of shear rate. Many examples are known in most of the existing procedures for solving engineering problems in food industry. They include the case study of processing concentrated fruit juices with suspended pulp particles, dairy cream, more or less concentrated tomato puree, apple puree, butter, minced meat, and infant foods. Among the food products that exhibit pseudoplastic behavior are all of which contain soluble high-molecular-weight substances and insoluble matter. Products that contain crystals and other particles like fat globules dispersed in a liquid phase, e.g., molten chocolate and ice cream mix have a yield stress. The apparent viscosity for a pseudoplastic material is $\eta_{app} = k \cdot D^{s-1}$.

There are many pseudoplastic food products that display more complex rheological behavior and with a yield stress that can be characterized in two ways, either by an extension of the power-law rheological equation of Herschel-Bulkley:

$$\tau = k \cdot D_s + \tau_0 \qquad (7.6)$$

or by an equation developed by Casson:

$$\tau^{1/2} = k_0 + k_i \cdot D^{1/2} \qquad (7.7)$$

where

k_0 = Casson yield stress
k_i = Casson plastic viscosity

Among this type of non-Newtonian materials are dilatant fluids. They show an increase in apparent viscosity with an increase in rate of shear and are not commonly found among liquid food products. They can be represented by the power law Equation (7.5), but in this case the flow behavior index, s, would be greater than 1 and less than infinity ($1 < s < \infty$). Such "shear thickening" is observed in materials with suspensions of solids at high solids content, when approaching the point of tightest packing. For example, corn-flour pastes can be dilatant.

There are other foods that possess the shear rate. Those food systems, when placed under steady flow at a constant rate, show a changing shear stress over time until an equilibrium value is achieved. Time-dependent behavior can be interpreted from viscoelasticity, thixotropy, or a combination of the two, and the power-law equation is not adequate for proper evaluation of such a system. Time-dependent materials can be subdivided into two classes—thixotropic and rheopectic. Time-dependent liquid foods for which the apparent viscosity decreases with time of shearing are known as thixotropic materials, whereas rheopectic fluids are those for which the apparent viscosity increases with time of shearing. In a thixotropic flow, the response to shear is instantaneous, and the time-dependent behavior can be observed when the shearing process continually alters the structure of the system. These structural changes include disentanglement of polymer molecules in solution and deflocculation of globules in emulsion. The rate of structure breakdown during shearing at a given rate depends on the number of linkages available for breaking and, therefore, decreases with time. In the case of rheopectic fluids, the structure builds up in shearing, and this phenomenon can be regarded as the reverse of thixotropy. In fact, rheopectic behavior is often referred to as antithixotropy. Thixotropy is generally defined as the continuous decrease of apparent viscosity with time under shear and subsequent recovery of viscosity when the flow is discontinued. The simplest example of such a system undergoing structural changes, known as symmetrically thixotropic fluid, is shown in Figure 7.8.

Once shearing is stopped (the external force is no longer acting), the rate of the system structural recovery is the same as the rate of structural breakdown under steady shear. In foods, the time needed for recovery may vary significantly from one thixotropic product to another. The time dependency can be observed in food systems such as concentrated emulsions, sols, and gels.

Semi-solid foods belong, in general, to a group that can be characterized by viscoelastic parameters. Viscoelasticity is due to delayed motion and retarded response of a system to a shear resulting from a joint viscous and elastic nature. Jellies and desserts fall into this category. Jams are also

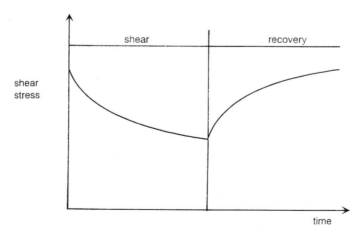

Figure 7.8 Flow behavior of a symmetrically thixotropic system. [Adapted from Harris (1977).]

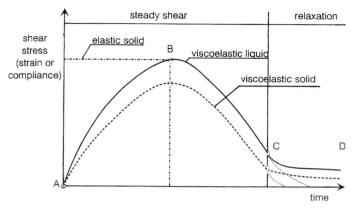

Figure 7.9 Flow behavior of a viscoelastic system. Region AB and CD represent viscoelastic behavior, BC represents structural breakdown during steady shear.

included; however, some additional measure of elasticity of the product is required.

Measurements of stress or viscosity decay at a constant or steady shear rate have also been used to characterize structural breakdown. The classical approach to characterizing structural breakdown is the measurement of the hysteresis loop [Figure 7.7(d,e)]. The area enclosed by the loop is the first indication of the degree of structural breakdown, and depends on the previous shear history and both the rate of change in shear rate and the maximum value of the shear rate. The relation between rheological properties and the food hysteresis loop area is a very complicated and complex shear history. Many food systems show viscoelastic behavior with time-dependent flow properties similar to that shown in Figure 7.9.

Among others, viscoelastic rheological behavior of such a food system can be characterized by a creep compliance test, where a constant stress is applied and the strain is followed as a function of time (t). Creep compliance is generally expressed in terms of the ratio, where strain (t) is over the stress. In this case, the stress relaxation is normally expressed in terms of ratio, where stress (t) is over a strain. For ideal elastic solid, the stress or strain will be independent of time, and the stress divided by the strain will be the elastic modulus of the material.

In the real food polymers, a distinction can be made between a viscoelastic solid, which contains some cross-links that are permanent, and a viscoelastic liquid where, under the influence of a stress, the relative movement of whole molecules will be observed. As is shown in Figure 7.9, in the case of a viscoelastic solid, after application of the stress, the strain will eventually reach a constant value, and upon removal of the stress, the strain will finally return to the remaining value of food primary energy, which was not entirely dissipated. For a viscoelastic liquid, a permanent deformation will remain after removal of the stress. In the stress relaxation area, the deformation value will decay to zero for a viscoelastic liquid, whereas for a solid, it will reach a constant nonzero value. It can also be seen either as a decreased value to the zero or constant nonzero value as pointed out by the dashed line. Both values characterize the rheology parameters of foods under certain conditions. One of the reasons for this is that the factors of time-dependent foods are not necessarily related to their elastic modulus. This can be explained by the series of small deformations without rupture, which are dependent in different ways and are based on the primary molecular microstructure of foods. The time required for the stress to relax to a definite fraction of its initial value is the relaxation time.

THE EFFECT OF COMPOSITION AND PROCESSING PARAMETERS ON THE RHEOLOGICAL BEHAVIOR OF FOOD SYSTEMS

ENERGY, STRESS, AND DEFORMATION

Stress and strain or deformation can be useful in microscopic or molecular descriptions of how the observed phenomena come about. They have directional properties, which distinguish an elongation from shear, for example. When the stress and strain may depend on time, it can be either the unchanging equilibrium state or a steady flow in which strain increases at a constant rate. The magnitude is also important; i.e., if doubling the strain leads to a doubling of the stress, then the behavior—because of the geometry and time patterns being identical—is called linear. In food industrial practice, nonlinear phenomena are mostly the rule, and the way rheological behavior actually appears depends on the time scale of the process in which it is observed. The ratio of a material's relaxation time to the time over which behavior is observed is called the Deborah number. As the Deborah number becomes smaller, the behavior changes from solid to fluid. In case of suspensions and dispersions of solid particles in a fluid, the effect of concentration, size, shape, and arrangement of the dispersed particles must also be considered.

In general, for the effect of composition and processing parameters of the food system, the external mechanical work needed to make a material deform or generate a flow pattern is used in three ways:

- It is stored reversibly so that the material can give back the energy as mechanical work.

- It is transformed to chemical energy in bonds or weak links between particles.
- It is dissipated into heat and lost.

Among all the deformations and flow patterns investigated in food science, one in particular is of predominant importance: steady shear flow. In this case, the geometry is shearing, the dependence on time is a steady increase in the shear, and the magnitude is arbitrary. The most common behavior in the food system is shear thinning or the pseudoplasticity curve, also known as a viscoelastic characteristic of materials such as polymers. In these, elasticity arises from the tendency of the polymer segments to take on random equilibrium arrangements due to their thermal motion.

A viscous contribution results from friction as one segment slides past another. The same friction also tends to drag the molecule out of its equilibrium shape. However, at low shear rates, the viscous stress is too small to do this, so that the shape of the molecule and also the food system's viscosity do not change. At larger shear rates, the viscous stress can deform the molecule into a shape past which the flow pattern occurs with more lowering of the viscosity. At a higher shear rate, the molecules cannot be deformed anymore. The shear rate can then be increased indefinitely without a further drop in viscosity. The equilibrium, considered as a configuration of the molecule, is restored when shearing stops, so that the deformation is a means of storing mechanical work recoverably.

Energy can also be stored by other ways on a microscopic scale, for example, by electrical charges being forced near each other in colloidal systems, as well as emulsion drops being distorted from the spherical shape. In this case, the surface tension gives them at rest or by stabilizing surfactant layers on dispersed particles being pressed into each other.

The mechanical work put into a food system with respect to the chemical transformation is typically associated with changes in the bonding between food components. In accordance with such changes, energy can be used to break links initially holding particles in chains or aggregates, leading eventually to all the particles being unbounded. Because the aggregation of particles causes a greater disturbance to the flow, such a system would be shear-thinning. Due to the passage of time, it would also be thixotropic when steady flow was started.

In other foods, the effect of composition and processing parameters can be achieved by different ways of storing work. It would be done by shear thickening. In this case, the particles stick together when they collide, and the link formation promoted by shearing is less likely. It would lead to dilatancy when steady-state viscosity is plotted against shear rate, and to rheopexy when an increase in viscosity takes place with time at the onset of flow. These phenomena occur only in highly concentrated dispersions of particles; i.e., randomly arranged particles interfere with each other greatly in flow conditions. If the interaction between particles allows layering within a certain time needed to accomplish this process, rearrangement of the food dispersed system will be more perfect in slower flows at low shear rates. The viscosity should then increase with shear rate. This final effect of composition and rheological behavior of such a food system is strongly related to the size of particles. If the particles are unequal in size, very small ones can fit into gaps between larger ones, allowing flow at a higher concentration than for equal particles. At the same concentration, a dispersion with a wide range of particle sizes has a lower viscosity than one with uniform particles. In some cases, the effect of food system composition regarding process parameters can be interpreted by catastrophic changes in component organization, i.e., antithixotropy, hysteresis, and relaxation period of time. Such an abrupt change can be observed, for instance, in shear thickening when a critical shear rate has been exceeded.

In general, the rheological properties of the food system have to be described in terms of particle sizes, shapes, surfaces, volumes, lengths, and their frequencies. Such structural characteristics of food systems are in reality the visual representations of highly ordered biological molecules. By combining such structural and functional data in equations that can then be integrated into mathematical and rheological models, opportunities are being created for analyzing complex food system responses and dissecting them into data of processing parameters and simpler, more interpretable food texture, quality, and functionality.

STRUCTURE DEVELOPMENT

A specific food product can be made by using a variety of recipes and processes, which differ in their demand on the functional ingredients. A detailed understanding of any effect of food system composition involves the following issues:

(1) Thermodynamic considerations: Is the reaction feasible to be carried out in the practice of food engineering? Whether or not a particular process is likely to take place can be ascertained by a theoretical consideration of the energy associated with the reaction. To what extent can it be calculated because of the enthalpy change, the free energy, and the entropy change, and to keep the equilibrium constant?

(2) Food system reaction kinetics: It is important to know, because of the final effect of food system composition, how fast the reaction is. It may be perfectly feasible from a thermodynamic point of view, yet it may be of no practical value because it takes place far too slowly.

(3) Reaction mechanisms: It is necessary to be familiar with what actually happens in all changes during a food engineering operation and to keep it in control because of the rheological parameters and their influence on food system composition. The fullest understanding of a reaction involves a study of the possible mechanisms by which one set of bonds is broken while another one is formed during the same process conditions.

(4) Separation of the product: A reaction mechanism intended to make one particular product will only be useful if it is relatively easy to separate that product from the mixture remaining at the end of the reaction. There are several possible results of physical separation: interruption of liquid bridging, lubrication, competition for adsorbed water, cancellation of electrostatic changes and molecular forces, and modification of crystalline lattice. All these results can influence flowability and reaction mechanisms in different food systems. They can also be used as a method for evaluation of flowability by the following attempt: direct flow rate measurement, angle of repose, shear and tensile strength measurements, the unconfined yield stress, angle of internal friction, cohesion, and plots of the whole flow function.

All of them are worthy in order to gain knowledge relevant to the processing parameters of non-Newtonian liquid and semi-solid food systems, particularly in the following areas:

- methodology adequate for the characterization of non-Newtonian materials with regard to changes taking place during shearing and heating
- relationships between flow properties and residual time in continuous processes
- better use of flow characteristics for the design of process equipment
- the effect of residual time for the quality of foods with regard to their microbiology, nutrition, and sensory and functional properties

Understanding the thermodynamics, reaction kinetics and mechanisms, and separation process, requires accurate analysis of confirmation and physical interaction in the absence of water or in the low-mixture environments.

There are some processes, like grinding for example, where the primary cell structure is changed significantly and can liberate the internal cell components. These liberated components can be purified, and their functional properties can be exploited to create foods whose textures are completely different from those of original raw materials, e.g., bread, biscuits, or sausages. Furthermore, by following this way, it is possible to use the natural texturizing properties of certain components. Such agents as emulsifiers, hydrocolloids, and proteins have opened up—through the emulsifying, whipping, softening, preventing crystallization, thickening, gelling and stabilizing facilities—new development possibilities in a wide range of food products, namely a few beverages, cream soup, sauce, salad dressing, bread, biscuits, pastry, sausages and meat spread, jam, gel, cream and whipped dessert, ice cream, and confectionery. These new agents, combined with improved processing parameters and preservation technologies, have led to the development of new foods with varied textures, such as those of liquids, semi-solids, solids, pastes, gels, and foams. Either single or several texturizing agents can often achieve the desired effect for food preparation, and the proper ingredients must be correctly selected and dosed, as well as incorporated at a precise processing stage in various processing systems.

GLASS TRANSITIONS AND STATE DIAGRAMS

A complete textural evaluation with respect to the processing parameters and rheological behavior of the food system may be logistically impossible. More importantly, not all these data may be required because some factors influencing the structural composition effect are probably multicorrelated parameters and more or less sensitive than others. The complex force-deformation and property frequency changes-response profiles of a variety of food materials are particularly amenable to fractal (Peleg, 1993) and to the phase glass transition. Food system composition effects can be characterized in terms of their fractal or noninteger dimension, as well as through the weight fraction of solids, when they are transformed into the viscous, liquid state. The application of fractals in image analysis (Kalab et al., 1995) and glass transition processes related to rheological changes in food systems with various aspects of food science, including texture evaluation, has only recently been addressed (Roos, 1995, Kokini et al., 1995). The common view of glass transition structure is that both short-range and long-range order are absent, and the structure formation process is best described as a continuous random network. Recently, it has been reported (Roos, 1995) that glass transition is a second-order phase transition. In some cases, the local ordering in food system composition prevails over an intermediate range of several atomic or molecular units. In the phase-transition process for gas-to-particle conversion, the following are included: homogeneous, homomolecular nucleation, i.e., the formation of a new, stable liquid or solid ultrafine particle from the gas involving one gaseous species only; homogeneous, heteromolecular nucleation, that is, the formation of a new particle involving two or more gaseous species, typically one of these being water; and heterogeneous, heteromolecular condensation, that is, the growth of preexisting particles due to deposition of molecules from the gas phase. Glass-phase transition processes result in exciting changes in free volume, molecular mobility, and physical properties of amorphous materials and can be detected from changes in mechanical, thermal, and dielectric properties of a food system. Knowledge of the relationships between glass transition temperature (T_g) and physiochemical changes is very important in predicting effects of composition on food rheological behavior in various processes, including, among others, agglomeration, baking, extrusion, dehydration, freezing, and storage conditions. The temperature of T_g of food system components can influence food properties, cause deteriorative changes, and can be useful for visualizing processing parameters through the use of state diagrams. These drawings can also be used to show the relationships among components of

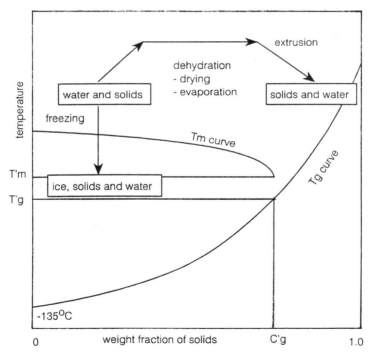

Figure 7.10 State diagram of water-plasticizable materials. Food solids can be transformed into a glassy state in processing, which removes water, e.g., extrusion, dehydration, and freezing. The glass transition temperature T_g decreases with increasing water content. Maximally, freeze-concentrated solids have glass transition at T_g', and the corresponding concentration of solids in the maximally freeze-concentrated phase C_g'. Equilibrium melting of ice occurs at T_m (from Roos, 1995).

a food system, temperature, and stability of food materials, as well as to predict changes under various process conditions (Figures 7.10 and 7.11). The case of cereal proteins in the prediction of material process phases is shown in Figure 7.12.

DYNAMICS MAP OF FOOD SYSTEM

The glass transition processes in foods may result from a rapid removal of water from solids. Based on that, for instance, the T_g values of anhydrous polysaccharides are high, and the food materials may decompose at temperatures below T_g (Kokini et al., 1994; Roos and Karel, 1991b). The glass transition temperature affects viscosity, stickiness, crispness, collapse, and crystallization, as well as ice formation, and can strongly influence deteriorative reaction rates. This provides a new theoretical and

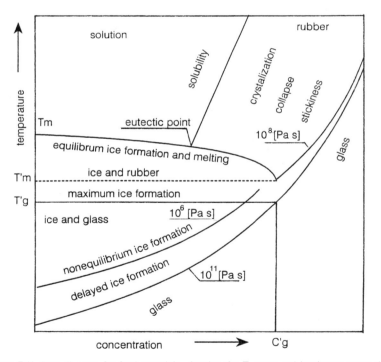

Figure 7.11 State diagram for food materials, showing the T_g curve and isoviscous states above T_g. Maximally, freeze-concentrated solids with a solute concentration of C_g' have T_g at T_g'. Ice melting within a maximally freeze-concentrated material occurs at T_m'. Equilibrium melting curve shows the equilibrium melting point T_m as a function of concentration (from Roos and Karel, 1991b).

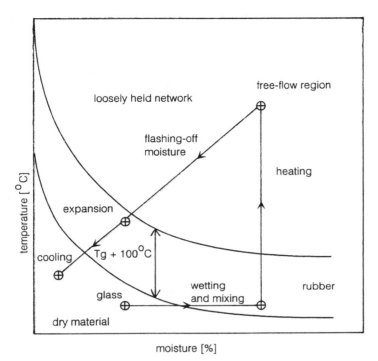

Figure 7.12 Transformation in cereal proteins during the wetting, heating, and cooling/drying stages of extrusion cooking, as seen on a hypothetical phase diagram (from Kokini et al., 1994).

experimental framework for the study of food systems, to unify structural and functional aspects of foods, described in terms of "water dynamics" and "glass dynamics."

The term *water dynamics* indicates the mobility of the plasticizing diluent and a theoretical approach to understanding how to control the water movement in glass-forming food systems. The term *glass dynamics* deals with the time- and temperature-dependence of relationships among composition, structure, and thermomechanical properties, as well as the functional behavior of food systems.

The functional aspects in terms of water dynamics and glass dynamics and the appropriate kinetic description of the nonequilibrium thermomechanical behavior of food systems has been illustrated as a "dynamics map," as shown in Figure 7.13.

The dynamics map represents both the equilibrium and nonequilibrium aspects. The equilibrium regions are described through two dimensions of temperature and composition. The major area of the dynamics map, shown in Figure 7.13, represents a nonequilibrium region of most far-reaching technological consequence to aqueous food systems. The non-equilibrium regions require for their description the third dimension of time, expressed as t/τ, where τ is a relaxation time. Nonequilibrium physical states determine the time-dependent thermomechanical, rheological, and textural properties of food systems. Based on the WLF (Williams-Landel-Ferry) mechanism, the glass transition as the food's reference state can be concluded and identified from the dynamics map. The dynamics map can be used to estimate the mobility transformation in water-compatible food polymer systems in terms of the critical variables of time, temperature, moisture content, and pressure.

The glass transition as the reference state can be used to explain all transformation in time, temperature, and structure composition effects between different relaxation states for technologically practical food systems in their nonequilibrium nature. Among others, there are specific examples like reduced activity and shelf-stability of freeze-dried proteins and living

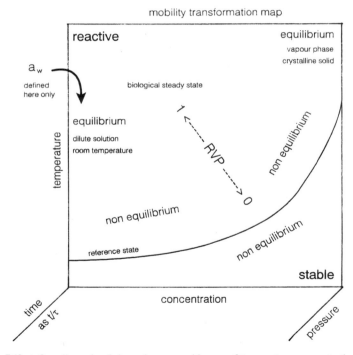

Figure 7.13 A four-dimensional dynamics map, with axes of temperature, concentration, time, and pressure, which can be used to describe mobility transformation in nonequilibrium glassy and rubbery systems, where a_w—water activity, RVP—relative vapour pressure (from Slade and Levin, 1991).

cells, graininess and iciness in ice cream, lumping of dry powder, bloom on chocolate, recipe requirements for gelatin desserts, cooking of cereals and grains, expansion of bread during baking, collapse of cake during baking, cookie baking effects of flour and sugar, and staling of baked products. Glass dynamics has been found as a very useful concept for elucidating the physiochemical mechanisms of structural/mechanical changes in various melting and (re)crystallization processes, including, for example, the gelatinization and retrogradation of starches. Glass dynamics can also be used to describe the viscoelastic behavior of amorphous polymeric network-forming proteins such as gluten and elastin (Slade and Levin, 1991).

These unified concepts, based on water dynamics and glass dynamics, have been used to explain and predict the functional and rheological properties of food systems during processing and their effects on time-dependent structural and mechanical factors related to quality and storage stability of food microstructure.

THE IMPORTANCE OF THE RHEOLOGICAL PROPERTIES OF FOODS FOR PROCESS DESIGN AND CONTROL

Accurate and reliable rheological data are necessary for the design and control of fluid-moving machines, sizing of pumps and other transport processes, to estimate velocity, shear and residence-time distribution in continuous mixing, and evaluate heating rates. Temperature differences occur in heat processing of foods. The nonuniform distribution of heat is due to the fact that the heat field intensity is not homogeneous over the entire object of application. It depends on the coefficient of retention, heat absorption, and the heat resistance at the walls of the equipment.

For many foods, the temperature effects are related to the changes in apparent viscosity:

$$\ln \eta_{pp} a = a + b/T \qquad (7.8)$$

where

η_{pp} = apparent viscosity
T = temperature in degrees Kelvin
a, b = constants

Illustrating the rheological behavior and its complex nature of molten chocolate will be considered. Chocolate is a suspension of solid particles in a fluid medium. The main solid particles in chocolate are cocoa and sugar, and the fluid is cocoa butter. There are other minor ingredients such as surface active agents, water, milk solids, and butter fat. Cocoa butter is Newtonian and time-dependent. Large quantities of solid particles, as well as surface and moisture active agents, can change the flow behavior pattern of molten chocolate to a non-Newtonian flow. A typical flow curve for molten chocolate has a yield stress, and the apparent viscosity falls rapidly with increasing rate of shear (Figure 7.14). Many factors influence the flow behavior of chocolate. The most important ones are fat content, type and quantity of surface active agent, moisture content, temperature, degree of shearing, and particle size and its distribution. The latter, in accordance with quality parameters and consumer acceptance, should not contain cocoa particles larger than 15 μm in diameter. Molten chocolate is, in many cases, time-dependent, which can be expressed by the form:

$$\tau = k \cdot t^{-m} \tag{7.9}$$

where

τ = shear stress
t = time
k = constant
m = index of time dependency

Chocolate can be characterized by a yield stress and plastic viscosity, i.e.,

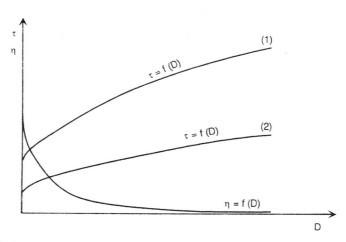

Figure 7.14 Relationship among shear stress, viscosity, and shear rate as flow behavior of molten chocolate: (1) temperature = 35°C, (2) temperature = 45°C (arbitrary scale).

as the Bingham plot. Another curve was established by Casson (reported by Holdsworth, 1971) in which chocolate is characterized by the yield value and plastic viscosity. The Bingham's plot is mainly used for process design and its control in production of plain chocolate. In the case of Casson plots, some molten chocolates, particularly those containing surface active agents, did not give a straight line relationship. To overcome this difficulty another expression was developed (Elson, 1977):

$$\tau = \eta_p \cdot D + B \cdot \sinh^{-1} D + \tau_0 \qquad (7.10)$$

where

τ = shear stress
D = rate of shear
η_p = plastic viscosity
τ_0 = yield stress
B = interaction effect

The flow behavior of molten chocolate can also be affected by changes in processing conditions. These may lead to different values because of the effect of non-Newtonian flow at the walls of the processing equipment. The wall shear rate is often characterized by the Nusselt number equation, which contains the so-called δ factor. The δ factor is the ratio of the wall shear rate for a non-Newtonian fluid to that of a Newtonian fluid at the same flow rate. For power law fluids, this factor can be calculated using the following formula:

$$\delta = (3n + 1)/4n \qquad (7.11)$$

where n = flow behavior index. The temperature-dependent properties of molten chocolate would have a major effect on heat transfer. Some assumptions are made that, during many heating operations, the temperature-dependent effects are more relevant than the degree of pseudoplasticity of the fluid. Solutions are often obtained in cases where heat is generated due to viscous dissipation of energy and by phase transition changes. The prediction of temperature and velocity profiles are also very important in a wide range of different heat transfer situations. Attempts to calculate and to predict the heat transfer rates in process design, for example, in agitated vessels, are complicated by the difficulties in accounting for the geometry of the system and by the complex rheological behavior of most liquids and semi-solid foods. Heat transfer to both Newtonian and non-Newtonian food systems and the prediction of the rate of heat transfer in both jacketed vessels and vessels containing heating coils, for agitation produced by paddles, turbines, propellers and anchors, can be considered by general types of correlation of the form

$$\text{Nu} = f(\text{Re, Pr, Vi, dimensionless geometrical factor}) \qquad (7.12)$$

where

Nu = Nusselt number
Re = Reynolds number
Pr = Prandtl number
Vi = η/η_w is the viscosity ratio
η_w = viscosity at the wall of equipment

For most non-Newtonian food systems, it is obvious that, because the shear rate throughout the material in the vessel varies, so does the apparent viscosity. This leads to a problem in specifying the viscosity, which is to be used in the Re, Pr, and Vi numbers, when this type of equation [Equation (7.12)] is applied in process design and control. The prediction of heat transfer coefficients in equipment handling non-Newtonian food products based on knowledge of the flow curve and its dependency changes for which it will be encountered is a prerequisite to any process design and its control considerations.

REFERENCES

Augilera, J. M. 1995. Gelation of whey proteins. *Food Technol.*, 49(10):83–89.

Elson, C. R. 1977. Increased design efficiency through improved product characterisation. *Symp. of Chemical Engineers,* May 10–11, 1977, West South Branch, U.K., pp. 96–104.

Gorris, L. M. G. 1995. Food preservation by combined processes. *Proceedings of the 1st Main Meeting "Copernicus Programme."* Dec. 6–9, 1995, Porto, Portugal.

Harris, J. 1977. *Rheology and Non-Newtonian Flow.* Longman Group, Ltd., London.

Holdsworth, S. D. 1971. Applicability of rheological models to the interpretation of flow and processing behaviour of fluid food products. *J. Texture Stud.*, 2:393–418.

Kalab, M., Alan-Wojtas, P. and S. S. Miller, 1995. Microscopy and other imaging techniques in food structure analysis. *Trends in Food Sci. and Technol.*, 6(60):177–186.

Kokini, J. L., Eads, T. and R. D. Ludescher, 1993. Research needs on the molecular basis for food functionality. *Food Technol.*, 47(3):36S–42S.

Kokini, J. L., Cocero, A. M. and M. Madeka, 1995. State diagrams help predict rheology of cereal proteins. *Food Technol.*, 49(10):74–82.

Kokini, J. L., Cocero, A. M., Madeka, H. and E. de Graaf, 1994. The development of state diagrams for cereal proteins. *Trends in Food Sci. and Technol.*, 5(50):281–288.

Mitchell, J. R. 1980. The rheology of gels. *J. Texture Stud.*, 11:315–337.

Matuszek, T. S. 1995. Raw materials and food processing with regard to the predictive microstructure. Abstracts, *9th World Congress of Food Sci. and Tech.*, July 30–August 4, 1995, Budapest, Hungary, 2, p. 136.

Peleg, M. 1977. Contact and fractures as components of the rheological memory of solid foods. *J. Texture Stud.*, 3:194–205.

Peleg, M. and E. B. Bagley, 1993. *Physical Properties of Foods*. AVI Publish. Co., Westport, CT.

Peleg, M. and A. M. Hollenbach, 1984. Flow conditioners and anticaking agents. *Food Technol.*, 38(3):93–102.

Peleg, M. 1993. On the use of the WLF model in polymers and foods. *Crit. Rev. Food Science Nutrition*, 32:59–66.

Roos, Y. H. and M. Karel, 1991a. Plasticizing effect of water on thermal behaviour and crystallisation of amorphous food models. *J. Food Sci.*, 56:38–43.

Roos, Y. H. and M. Karel, 1991b. Applying state diagrams in food processing and product development. *Food Technol.*, 45(12):66–71, 107.

Roos, Y. H. 1995. Glass transition-related physiochemical changes in foods. *Food Technology*, 49(10):97–102.

Shoemaker, C. F. and P. I. Figoni, 1984. Time-dependent rheological behaviour of foods. *Food Technology*, 38(3):112–121.

Slade, L. and H. Levin. 1991. Beyond water activity. Recent advances based on an alternative approach to the assessment of food quality and safety. *Crit. Rev. Food Sci. Nutr.*, 30: 115–360.

CHAPTER 8

Food Colorants

JADWIGA WILSKA-JESZKA

INTRODUCTION

COLOR is an important quality aspect of both unprocessed and manu-factured foods. The natural colorants are unstable; thus, color of food products may provide an indication of biochemical and chemical changes during processing and storage.

However, color cannot be studied without considering the human sensory system. Perception of color is related to three factors: spectral composition of the light source, physical object characteristics, and eye sensitivity. Fortunately, the characteristic of the human eye for viewing color is fairly uniform, and it is not difficult to replace the eye by some instrumental sensor or photocell.

The color of food is the result of the presence of natural pigments or of added dyes. Natural pigments are generally considered the pigments occurring in unprocessed food, as well as those that can be formed upon heating, processing, or storage. These pigments can be divided into five groups:

- carotenoids—isoprenoid derivatives
- chlorophylls and hemes—porphyrin pigments
- anthocyanins—2-phenylbenzopyrylium derivatives
- miscellaneous naturally occurring colorants such as betalains, cochineal, riboflavin, and curcumin
- melanoidins and caramels—formed during food heating and storage

All those natural pigments are unstable and participate in different reactions, and so the food color is strongly dependent on conditions.

191

The use of synthetic organic colors has been recognized for many years as the most reliable and economical method of restoring some of the food's original shade to the processed product. An even more important application of synthetic dyes is using them for improving and standardizing the appearance of those products that have little or no natural color present, such as dessert powders, table jellies, ice, and sugar confectionery. The synthetic organic colors are superior to the natural pigments in tinctorial power, range and brilliance of shade, stability, ease of application, and cost-effectiveness. However, from a health safety viewpoint, they are not accepted by the consumers, so over the last ten years, increasing interest in natural food colorants has been observed.

NATURAL PIGMENTS

CAROTENOIDS

Structure

Carotenoids are widely distributed yellow natural pigments, occurring in leaves, together with chlorophyll, and in other parts of plants: roots, flowers, fruits, and grains. The name *carotenoids* has been derived from the major pigments of carrot (*Daucus carote* L.). The animals do not synthesize carotenoids, but they may change ingested pigments into animal carotenoids, e.g., in salmon and eggs.

Carotenoid molecules have a basic structure made up of eight isoprenoid residues, arranged as if two twenty-carbon units formed by head to tail condensation of four isoprenoid units connected tail to tail (Formula 8.1). The carotenoids comprise two groups of pigments: the carotenes, which are hydrocarbons, and the xanthophylls, which contain oxygen in the form of hydroxyl, metoxyl, carboxyl, keto, or epoxy groups. For designing the carotenoids that are derived by loss of a structural element by degradation, the prefix *apo* is used. Formulae 8.2–8.9 present the structures of the most important carotenoids.

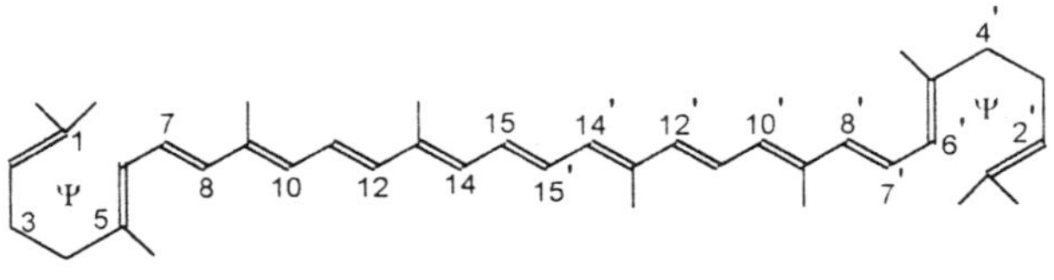

(8.1)

Lykopene

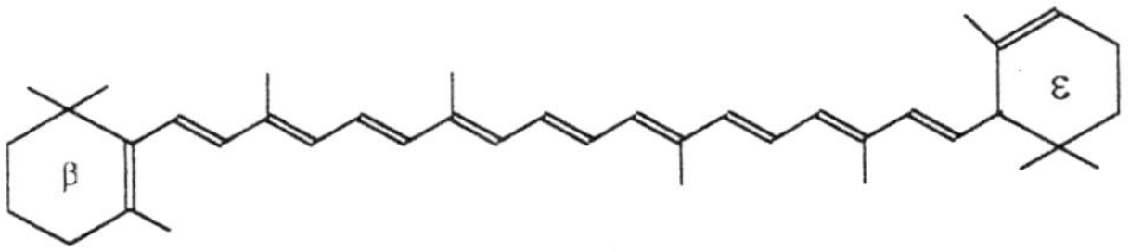

(8.2)

α – carotene

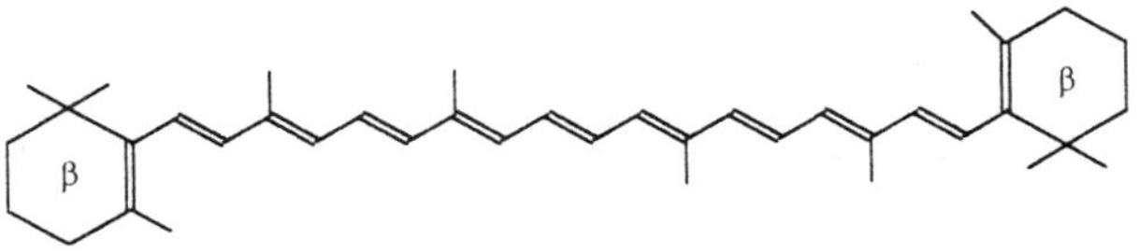

(8.3)

β – carotene

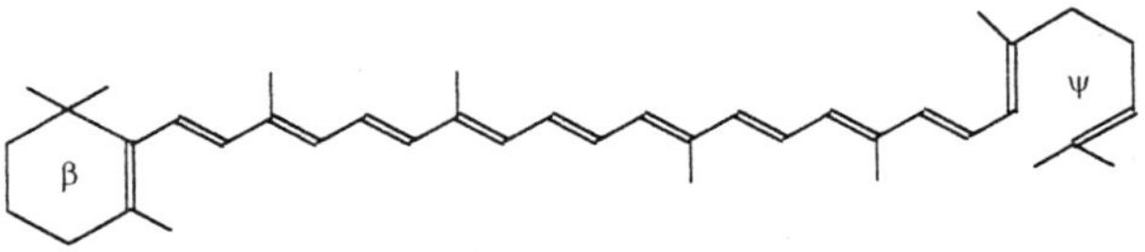

(8.4)

γ – carotene

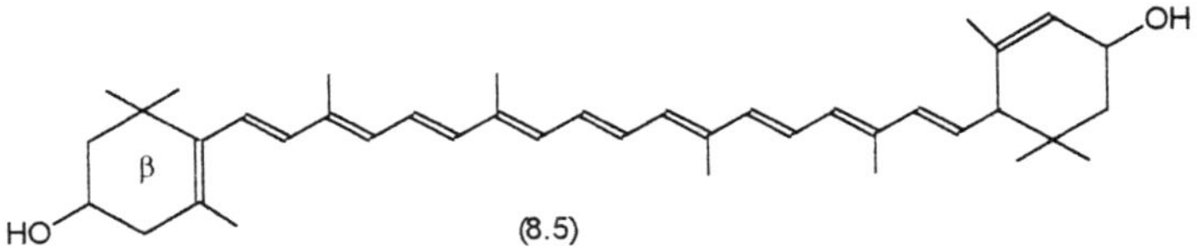

(8.5)

Lutein

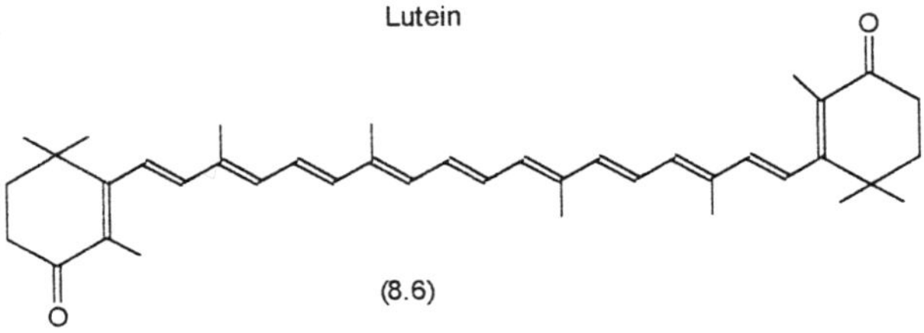

(8.6)

Canthaxanthin

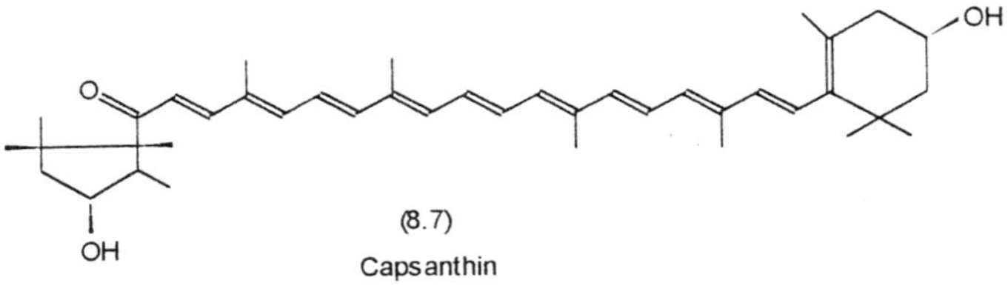

(8.7)

Capsanthin

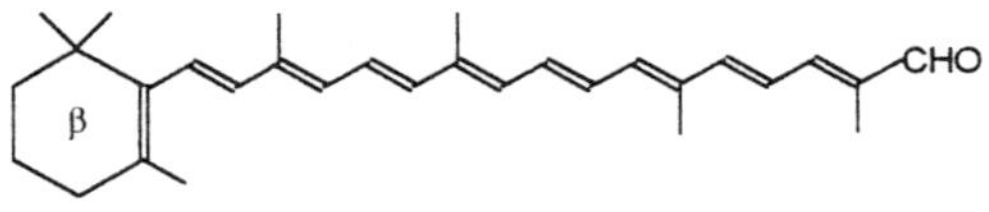

(8.8)

β – Apo - 8' - carotenal

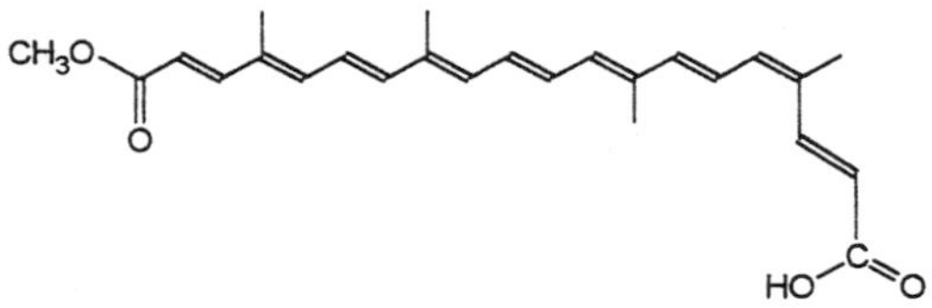

(8.9)

Bixin

The color is the result of the presence of a system of conjugated double bonds. A minimum of seven conjugated double bonds are required for the yellow color to appear. The increase in number of double bonds results in a shift of the major adsorption bands to the longer wavelengths, and the hue of carotenoids becomes more red.

In unprocessed plants, usually all-*trans* double bonds configuration occurs. Processing and storage can cause isomerization of carotenoids in foods and affect a change of color. The compounds with all-*trans* configuration have the deepest color. Increasing numbers of *cis* bonds result in gradual lightening of the color. The prefix *neo* is used for stereoisomers with one *cis* bond and the prefix *pro* for poly-*cis* carotenoids.

Occurrence

Carotenoids may occur in foods as simple mixtures of a few compounds or as very complex mixtures. The simplest mixtures usually exist in animal products, because the animal organism is limited in its ability to absorb and deposit carotenoids. Some of the most complex carotenoid mixtures are found in citrus fruits.

The three main carotenoids of green leaves are lutein, violoxanthin, and neoxanthin; others are produced in smaller quantities. In fruits, as the chlorophyll breaks down during ripening, large amounts of carotenoids are formed. Common carotenoids in fruits are α-carotene, β-carotene, lycopene, and xanthophylls. The last are usually present in esterified form. Some of the carotenoids, such as β-carotene, lycopene, and zeaxanthin, are very widely distributed and so become important as food components. However, the content of carotenoids usually does not exceed 0.1% of dry weight. Oxygen, but not light, is required for carotenoid synthesis.

In fruits and vegetables, β-carotene content is used as a measure of the provitamin A content. One mole of β-carotene can theoretically be converted, by cleavage of C 15 = C 15′ double bond, to yield two moles of retinal (Reaction 8.1). However, the physiological efficiency of this process appears to be only 50%. The observed average efficiency of intestinal β-carotene absorption is only two-thirds of the total content. Thus, a factor of one-sixth is used to calculate the Retinol Equivalent (RE) from β-carotene, and one-twelfth from the other provitamin A carotenoids in food (Combs, 1992).

Carrots are one of the best sources of β-carotene. The total carotene content in carrots is 50–100 ppm, consisting mainly of β-carotene, about 80%. Canning of carrots results in a 7–12% loss of provitamin A activity, because of *cis-trans* isomerization of α- and β-carotene. In dehydrated carrots, carotene oxidation and off-flavor development have been correlated. Corn contains one-third of carotenoids as carotenes, mainly β-carotene, and two-thirds as xanthophylls: lutein, zeaxanthin, and cryptoxanthin.

Crude vegetable oils contain carotenoids. Bleaching and hydrogenation leads to almost complete degradation of this compound. Particularly rich

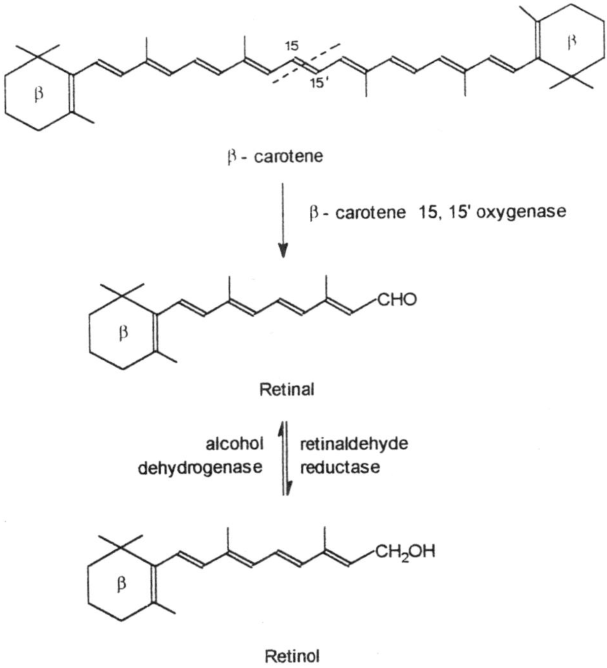

β - carotene

β - carotene 15, 15′ oxygenase

Retinal

alcohol dehydrogenase | retinaldehyde reductase

Retinol

Reaction 8.1 Formation of vitamin A from β-carotene.

in carotenoids (0.05–0.2%) is crude palm oil, containing mainly α-carotene and β-carotene. Egg yolk contains only xanthophylls: lutein, zeaxanthin, and cryptoxanthin (3–80 ppm). The muscle oil of redfish contains astaxanthin, lutein, and taraxanthin.

Carotenoids Used as Food Colorants

The most commonly used natural carotenoid extracts for foodstuffs are annatto, paprika, and saffron. Many other sources, including alfalfa, carrot, tomato, citrus peel, and palm oil, are also utilized.

Annatto [E 160(b)] is the orange-yellow, oil-soluble natural pigment extracted from the pericarp of seed of the *Bixa orellana* L. tree. The major coloring component of this extract is the diapocarotenoid bixin (Formula 8.9). Several other pigments, chiefly degradation products of bixin, are also present, including *trans*-bixin, norbixin, and *trans*-norbixin. Bixin is a methyl ester of a dibasic fatty acid, which on treatment with alkalis, is hydrolyzed to water-soluble norbixin. Two types of annatto are therefore available: an oil-soluble extract containing bixin and a water-soluble extract containing norbixin.

Paprika oleoresin [E 160(c)] is the orange-red, oil-soluble extract from sweet red peppers *Capsicum annum*. The major coloring compounds are xanthophylls: capsanthin (Formula 8.7), capsorubin as their dilaurate esters, and β-carotene. The presence of characteristic flavoring and spicy-pungency components limits application of this extract in foodstuffs.

Saffron—extract of flowers of *Crocus sativus*—contains the water-soluble pigment crocin, the digentiobioside of apocarotenic acid, crocetin, as well as zeaxanthin, β-carotene, and characteristic flavoring compounds. The yellow color of this pigment is attractive in beverages, cakes, and other bakery products; however, use of this colorant is restricted by its high price.

Carrot extracts [E 160(a)], carrot oil as well as palm oil and related plant extracts are also available on the market. Their main components are α- and β-carotenes (Formulae 8.2 and 8.3, respectively). Processes for the commercial extraction of carotene from carrots were developed. Purified crystalline products contain 20% α- and 80% β-carotene and may be used for coloring fat-based products as dispersion of microcrystals in oil.

Individual carotenoid compounds—β-carotene (Formula 8.3), β-apo-8′ carotenal (Formula 8.8), apocarotenoic ethyl ester, and canthaxanthine (Formula 8.6)—are synthesized for use as food colorants for edible fats and oils. These carotenoids, in combination with surface active agents, are also available as microemulsions for coloring foods with a high water content.

Properties

Carotenoid pigments are generally insoluble in aqueous media, have a low solubility in oils, and have a rather low rate of dissolution, particularly in pure crystalline state. The more important physical properties of four pure synthesized carotenoids are given in Table 8.1.

Carotenoids are highly sensitive to oxygen and light. Oxidation is accelerated by radicals and peroxides occurring in food as a result of lipid autoxidation, particularly where metallic catalysts such as copper, iron, and manganese are present. The hydroperoxides can directly attack carotenoids. Lipoxygenase involved in the decay of vegetable matter may also cause the destruction of carotenoids. Antioxidants, including ascorbic acid and their derivatives, tocopherols and polyphenolics, are used to suppress this oxidative degradation.

Processing and storage, depending on the condition, can cause isomerization and degradation of carotenoids in food. The color intensity and biological activity of *cis* isomers is less than that of all-*trans* molecules. The bipotency of *cis* isomers is no more than 50% of that of all-*trans* β-carotene. The result of oxidative degradation of carotenoid may be observed as changes in coloration of dehydrated carrots or paprika.

Due to oxidative degradation of carotenoids, aroma compounds are also formed, including β-ionone with a threshold value of 14 ppb in water. The formation of β-ionone in dehydrated carrots causes the undesired off-flavor, "odor of violets." Unsaturated ketones derived from carotenoids degradation are readily further oxidized. The stability of carotenoids in frozen and heat-sterilized foods is quite good, but poor in dehydrated products, unless the products are packaged in inert gas. Dehydrated carrots fade rapidly.

TABLE 8.1. Properties of Carotenoids Used as Food Colorants.

Carotenoid	Color	Solubility (g/100 ml), 20°C		λ_{max}	Vitamin A Activity (IU/mg)
		Oils	Ethanol		
β-Carotene	Yellow	0.05–0.08	0.01	455–456	1.67
Apocarotenoic ester	Yellow to orange	0.7	0.1	448–450	1.2
Apocarotenal	Orange to red	0.7–1.5	0.1	460–462	1.2
Canthaxanthin	Red	0.005	0.01	468–472	0

After Klaui H. and Bauernfeind Ch., 1981.

CHLOROPHYLL

Chlorophyll is the most widely distributed natural pigment. In living plant tissues, chlorophyll is present in colloidal suspension in chloroplast cells, in the form associated with protein and carbohydrates. The chlorophyll pigments are the same in all plants. Apparent differences in color are due to the presence of other associated pigments, in particular, xanthophylls and carotenes, which always accompany the chlorophylls. Typical leaf material contains about 2.5 mg/g total chlorophylls, 0.3 mg/g xanthophylls, and 0.15 mg/g carotenes (Humphrey, 1980). In many fruits, chlorophyll is present in the unripe state and gradually disappears during ripening as the yellow and red carotenoids take over.

The chlorophylls are tetrapyrrole pigments in which the porphyrin ring is in the dihydro form and the central metal atom is magnesium (Formula 8.10). There are two chlorophylls: a (blue-green) and b (yellow-green), occurring together in a ratio of about 3:1. Chlorophyll b differs from chlorophyll a in that the methyl group on carbon 3 is replaced with an aldehyde group. Chlorophyll is a diester: one group is esterified with methanol and the other with phytyl alcohol.

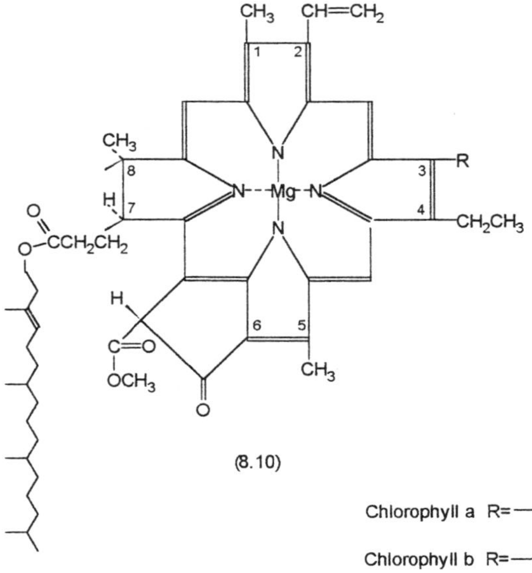

(8.10)

Chlorophyll a R=—CH$_3$

Chlorophyll b R=—CHO

The important chemical characteristics of the chlorophylls are

- the easy loss of magnesium in dilute acids or replacement by other divalent metals

- the hydrolysis of the phytyl ester in dilute alkalis or transesterification by lower alcohols
- the hydrolysis of the methyl ester and cleaving the isocyclic ring in stronger alkalis

Removal of magnesium gives olive-brown pheophytin a and b. Replacing magnesium by iron or tin ions yields grayish-brown compounds, while copper or zinc ions retain the green color. Upon removal of the phytyl group by hydrolysis in dilute alkali or by the action of chlorophyllase, green chlorophyllins are formed. Removal of magnesium and phytyl group results in pheophorbid formation (Figure 8.1).

Chlorophylls and pheophytins are lipophyllic due to the presence of the phytyl group, while chlorophyllins and pheophorbids without phytyl are hydrophyllic. The copper complexes of both pheophytin and pheophorbid have the metal firmly bound; it is not liberated even by the action of concentrated hydrochloric acid and not removed to any appreciable extent on metabolism; thus, it is acceptable for the coloration of foodstuffs. Both coppered and uncoppered chlorophylls and their derivatives are available as food colorants.

The oil-soluble chlorophylls (uncoppered and coppered pheophytin) are not widely used for food coloring because commercial purification has not proven to be as satisfactory as those for the water-soluble derivatives. Their stability is good towards light and heat but poor to acid and alkaline condition. Applications are found in canned products and confectionery on the levels 0.5–1 g/kg (Coulson, 1980).

The water-soluble chlorophyllins (uncoppered and coppered sodium or potassium pheophorbide) have good stability towards light and heat and moderate stability to both acid and alkalis. Food color usage is in canned products, confectionery, soups, and dairy products.

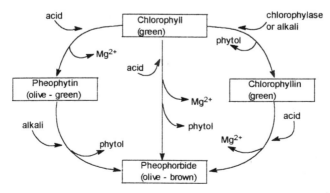

Figure 8.1 Reactions of chlorophyll.

HEME PIGMENTS

The color of meat is the result of the presence of two pigments: myoglobin and hemoglobin. In both pigments, the heme group is composed of the porphyrin ring system and the central iron atom bound with globin. In myoglobin, the protein portion has a molecular weight of 17,000 and in hemoglobin, about 67,000.

The central iron atom has six coordination bonds, each representing an electron pair accepted by the iron from five nitrogen atoms, four from the porphyrin ring, and one from a histidyl residue of the globin. The sixth bond is available for joining with any atom that has an electron pair to donate, e.g., O_2 or NO. The oxidation state of the iron atom and physical state of globin play an important role in meat color formation.

In the fresh meat, in the presence of oxygen, the reversible reaction of myoglobin (Mb) with oxygen occurs and as a result, bright red-colored oxymyoglobin (MbO_2) is formed. In both pigments, the iron is in ferrous form, and upon oxidation to a ferric state, the brownish metmyoglobin (MMb) is formed.

$$
\begin{array}{ccccc}
MbO_2 & \rightleftharpoons & Mb & \rightleftharpoons & MMb \\
\text{red} & & \text{purplish red} & & \text{brownish}
\end{array}
\qquad \text{(Reaction 8.2)}
$$

The oxymyoglobin and myoglobin exist in a state of equilibrium with oxygen. The ratio of these pigments depends on oxygen pressure. In meat, there is a slow oxidation of the heme pigments to metmyoglobin, which cannot bind oxygen.

Heating of meat results in the denaturation of the globin linked to iron, as well as in the oxidation of iron to the ferric state, and formation of different brown heme pigments named hemichrome.

In the curing of meats, the heme pigments react with nitrite of the curing mixture, and the red-colored, nitrite-heme complex nitrosylmyoglobin is formed but it is not particularly stable. More stable is the pigment with denatured globin portion, named nitrosylhemochromogen, found in meat heated in temperatures $> 65°C$. The reactions of heme pigments in meat and meat products have been summarized in the scheme shown in Figure 8.2 (see also page 142).

In the presence of thiol compounds as reducing agents, in the reversible reaction, myoglobin may form a green sulfmyoglobin. Other reducing agents, e.g., ascorbate, lead to formation of cholemyoglobin. This reaction is irreversible.

A potential source of red and brown heme pigments is animal blood and its dehydrated protein extracts, which mainly consist of hemoglobin and may be used as red and brown colorings to meat products. However, in most countries, their usage as a food coloring agent is not permitted.

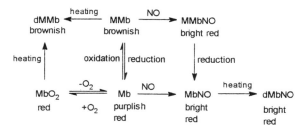

Figure 8.2 Reactions of myoglobin in meat. Mb – myoglobin, MMb – metmyoglobin, MbO₂ – oxymyoglobin, MbNO – nitrosylmyoglobin, dMMb – denatured metmyoglobin, dMbNO – denatured nitrosylmyoglobin.

ANTHOCYANINS

Occurrence and Structure

Anthocyanins are among the most important groups of plant pigments. They are present in almost all higher plants and are the dominant pigments in many fruits and flowers, giving them a red, violet, or blue color.

Anthocyanins are part of the very large and widespread group of plant constituents, known as flavonoids, which possess the same $C_6 - C_3 - C_6$ basic skeleton. They are glycosides of polyhydroxy and polymetoxy derivatives of 2-phenylbenzopyrylium or flavylium cation (Formula 8.11).

(8.11)

Flavylium cation

Differences between individual anthocyanins are the number of hydroxyl groups in the molecule; the degree of methylation of these hydroxyl groups; and the nature, number, and the position of glycosylation, as well as the nature and number of aromatic or aliphatic acids attached to the glucosyl residue. From about twenty known naturally occurring anthocyanidins, only six occur most frequently in plants, i.e., pelargonidin, cyanidin, peonidin, delphinidin, petunidin, and malvidin (Table 8.2). Substitution of the hydroxyl and methoxyl groups influences the color of the antho-

TABLE 8.2. The Most Important Anthocyanidins.

Anthocyanidins	R_3	R_5	λ_{max} (nm)	Color
Pelargonidin (Pg)	H	H	520	Orange
Cyanidin (Cy)	OH	H	535	Orange-red
Peonidin (Pn)	OCH_3	H	532	Orange-red
Delphinidin (Dp)	OH	OH	546	Bluish-red
Petunidin (Pt)	OCH_3	OH	543	Bluish-red
Malvidin (Mv)	OCH_3	OCH_3	542	Bluish-red

cyanins. Increase in the number of hydroxyl groups tends to deepen the color to a more bluish shade. An increase in the number of methoxyl groups increases redness. Because of the possibility of different ose and acid substitution at different positions, the number of anthocyanins is fifteen to twenty times greater than the number of anthocyanidins (Mazza and Miniati, 1993).

The ose molecules most commonly bonded to anthocyanidins are glucose, galactose, rhamnose, arabinose, and di- and trisaccharides such as rutinose, sambubiose, sophorose, or gentiobiose. Glycosylation of anthocyanidins almost always occurs in the 3-position, rarely in the 5- or 7-position.

Structural Transformation in Aqueous Media

According to Brouillard (1982), in acidic and neutral media, four anthocyanin structures exist in equilibrium: the flavylium cation (AH⁺), the quinonoidal base (A), the carbinol pseudobase (B), and the chalcone (C) (Reaction 8.3). The pH of the medium plays a particularly important role in the equilibrium between the different anthocyanin forms and, consequently, in color modification. Anthocyanins exhibit intense red coloration only in a very limited pH range, 1–3. This corresponds to an equilibrium between the red flavylium cation and the colorless carbinol base. An increase of pH causes reversible structural transformation with loss of red color. At a pH range of 4–6.5, flavylium salt is converted immediately to the two purple tautomeric neutral quinonoidal bases, which slowly hydrate to the carbinol pseudobase, with progressive disappearance of color. The reactions A → AH⁺, AH⁺ → B, and B → C are all endothermic; thus, any temperature rise strongly favors the ring-opened chalcon C at the expense of A, AH⁺, and B. Therefore in the higher temperature, chalcon content increases.

Color Stabilization and Transformation

The color of anthocyanins containing media depends on the structure

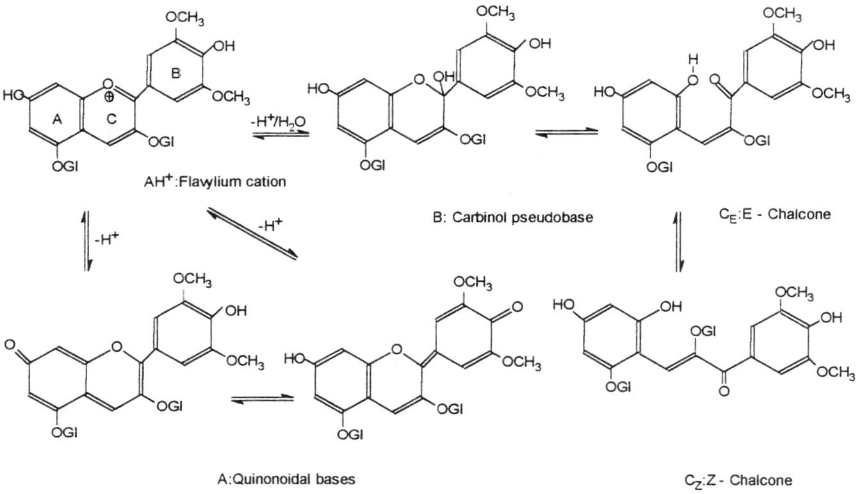

Reaction 8.3 Structural transformations of malvin in aqueous media.

and concentration of the pigment, pH, temperature, presence of copigments, metallic ions, phenoloxidase and β-glucosidase, ascorbic acid, sulfur dioxide, sugar and its degradation products, and other factors.

The structure of anthocyanin molecule has a marked effect on color intensity and stability. The increase of the number of hydroxyl groups in the B-ring shifts the absorption maximum to longer wavelengths, and the color changes from orange to bluish-red. Methoxyl groups replacing hydroxyl groups reverse this trend. The hydroxyl group at C-3 is particularly significant because it shifts the color from yellow-orange to red. The same hydroxyl, however, destabilizes the molecule, and the 3-deoxyanthocyanidins are much more stable than the other anthocyanidins. Similarly, the presence of a hydroxyl group at C-5 and substitution at C-4 both stabilize the colored forms. Glycosylation also affects the stability of this pigment, and the half-life of anthocyanidins is significantly lower than their corresponding 3-glucosides. Anthocyanins containing two or more aromatic acyl groups, such as cinerarin or zebrinin, are stable in neutral or weakly acidic media. This is possibly a result of hydrogen bonding between phenolic hydroxyl groups in anthocyanidins and aromatic acids. Brouillard (1981) and Goto et al. (1982) observed that diacylated anthocyanins are stabilized by sandwich-type stacking caused by hydrophobic interaction between the anthocyanidin ring and the two aromatic acyl groups.

The increase in anthocyanin concentration results in an increase in absorbance at λ_{max}, greater than expected according to the Beer-Lambert law. It is probably connected with anthocyanins self-association, the mechanism of which has not yet been fully described.

Intermolecular copigmentation of anthocyanins with other flavonoids, some phenolic acids, alkaloids, and other compounds, including anthocyanins themselves, increases the color intensity (hyperchromic effect) and a shift in the wavelength (batochromic shifts), giving purple to blue colors. The intensity of the copigmentation effect depends upon several factors, including type and concentration of both anthocyanins and copigments, pH, and temperature of the solvent (Brouillard et al., 1991). The pH value for maximum copigmentation effect is about 3.5 and may vary slightly depending on the pigment–copigment system. Color intensification by copigmentation increases with an increasing ratio of copigment to anthocyanins. Increasing the temperature strongly reduces the color-intensifying effect. The copigmentation phenomenon is widespread in nature and also occurs in fruits and vegetable products such as juices and wines.

Anthocyanins can form purplish-blue or slate-gray pigments with metals called lakes. This reaction may influence the color changes, if fruit products, during processing or storage, are in contact with metals such as tin, aluminum, or iron.

Stability of Anthocyanins Color

Anthocyanins appear to have low stability in all products manufactured from fruits, which limits the use of this pigment as food colorants. Two main groups of factors are responsible for the stability of anthocyanin's color in fruits during processing:

- initial composition of the fruits, with regard to anthocyanins and other constituents, including enzymatic systems
- physicochemical factors during fruits processing, such as temperature and light

Several enzymes, in particular, phenolases, peroxidases, and β-glucosidases, can spoil the quality of the initial product during the extraction of fruit juices or the preparation of processed products, leading to browning and to loss of color by enzymatic degradation of anthocyanins. Anthocyanins themselves are not good substrates for o-diphenol oxidase, but they are, instead, oxidized by chlorogenic acid → chlorogenoquinone redox shuttle phenomena. Thus, enzymatic oxidation of chlorogenic acid may be combined with nonenzymatic oxidation and polymerization of anthocyanins. The same phenomenon has been observed in the presence of catechins.

Ascorbic acid may have a protective effect with regard to anthocyanins since it reduces the o-quinones formed before their polymerization. However, ascorbic acid, as well as products of its degradation, increase the anthocyanin degradation rate.

Sulphur dioxide, widely used in fruit processing, at concentrations as low as 30 ppm, inhibits the enzymatic degradation of anthocyanins. At higher concentrations, it forms colorless SO_2-anthocyanin complex. This is a reversible reaction: after removal of sulphur dioxide, the color turns red again.

Regardless of the favorable action of high temperature on the blockage of enzymatic activities, anthocyanins are readily destroyed by heat during the processing and storage. A short time/high-temperature process was recommended for best pigment retention. For instance, in red fruit juices heated 12 min at 100°C, anthocyanin losses appear to be negligible.

The reaction listed above can proceed with different rates and can bring about different changes in color, depending on the composition of food products. Most frequently, the red color slowly turns brown. However, by correctly processing and storing fruits, color changes are so slow that they only slightly affect consumer appreciation of fruit products.

Anthocyanins (E 163) are available as spray-dried extracts of pomace remaining after wine or juice production from fruits, such as e.g., grapes, bilberry, or chokeberry. Anthocyanin concentrate may be used as food colorant at pH < 4.

BETALAINS

Betalains occur in centrosperme, mainly in red beet, and also in some mushrooms. They consist of red-violet betacyanins and yellow betaxantin. The major betacyanin is betanin (Formula 8.12), glucoside of betanidin, which accounts for 75–95% of total pigments of beets. The other red pigments are isobetanin (C-15 epimer of betanin), prebetanin, and isoprebetanin. The latter two are sulphate monoesters of betanin and isobetanin, respectively. The major yellow pigments are vulgaxanthin I and II (Formula 8.13). High betalains content in beetroot, on average 1% of the total solids, makes it a valuable source of the food colorant.

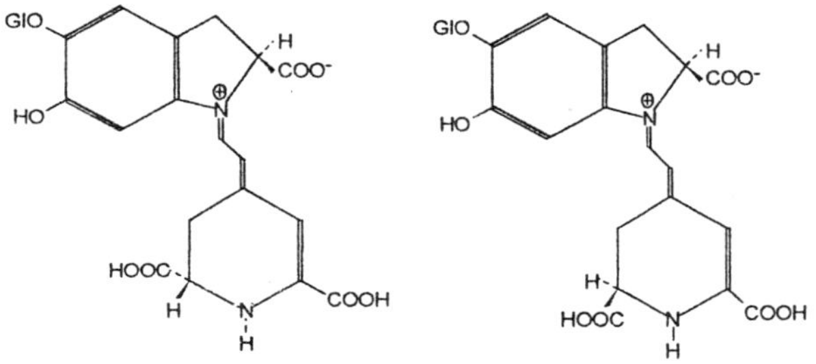

(8.12)

Betanin   Isobetanin

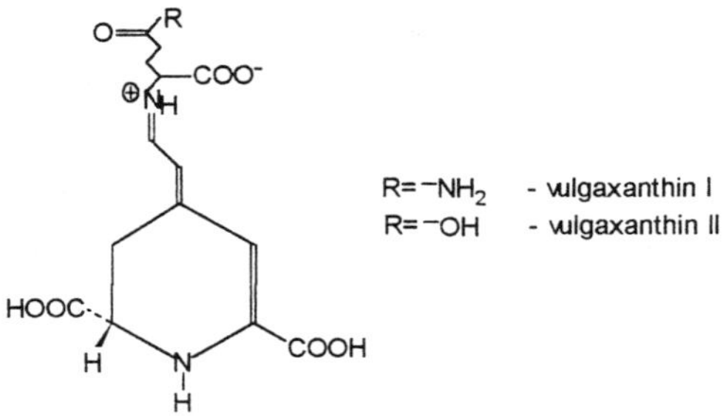

R= −NH₂ - vulgaxanthin I
R= −OH - vulgaxanthin II

(8.13)

The color stability of betanin solution is strongly influenced by pH and heating. Betanin is stable at pH 4–6, but thermostability is greatest at pH 4–5. As a result of betanin degradation, cyclodopa and betalamic acid are formed (Reaction 8.4). This reaction is reversible (Czapski, 1985). Light and air have a degrading effect on betanin. These effects are cumulative, but some protection may be offered by antioxidants such as ascorbic acid. Small amounts of metallic ions increase the rate of betanin degradation; therefore, a chelating agent can stabilize the color. Many protein systems present in food products also have some protective effect.

Beetroot Red (E 162), available as liquid beetroot concentrate and as beetroot concentrate powders, is suitable for products of relatively short shelf-life, which do not undergo as severe heat treatment, e.g., meat and soya protein products, ice cream, and gelatin desserts.

QUININOID PIGMENTS

These pigments are widely distributed. They are the major yellow, red, and brown coloring materials of roots, wood, and bark, and they also occur at high levels in certain insects. The largest group is that of anthra-

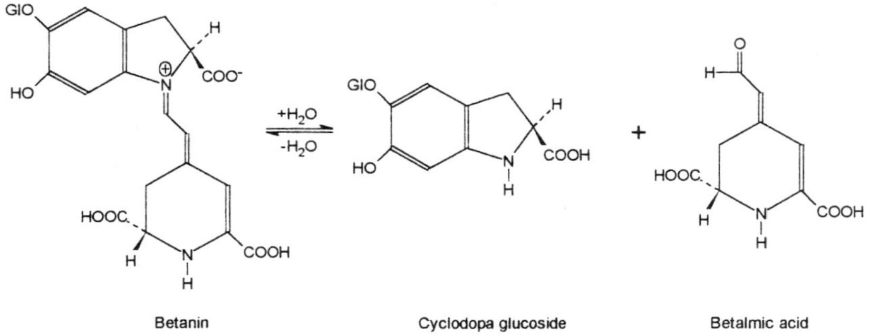

Betanin   Cyclodopa glucoside   Betalmic acid

Reaction 8.4 Degradation of betanin.

quinone pigments. The most important quininoid pigments, commercially available for use in foods, are cochineal and cochineal carmine.

Cochineal (E 120) is the red coloring matter extracted from the dried bodies of female insects of the species *Dactylopius coccus Costa* or *Coccus cacti* L. The insects are cultivated on the cactus plants in Peru, Equador, Guatemala, and Mexico.

The major pigment of cochineal is polyhydroxyanthraquinone C-glycoside (carminic acid) (Formula 8.14), which may be present at up to 20% of the dry weight of the mature insects. Cochineal extract or carminic acid is rarely used as coloring material for foodstuffs but is usually offered in the form of their lake. Aluminum complexes (lakes) can be prepared with various ratios of cochineal and alumina varying from 8:1 to 2:1, having corresponding shades from pale yellow to violet.

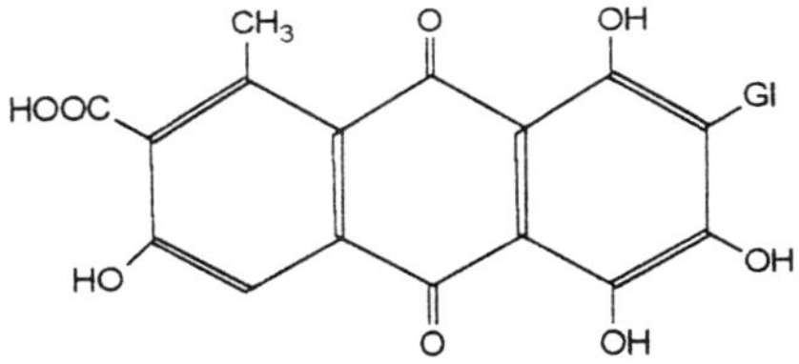

(8.14)

Carminic acid

Cochineal carmine is insoluble in cold water, dilute acids, and alcohol and slightly soluble in alkali, giving a purplish-red solution. The shade becomes more blue at higher pH. Cochineal carmine is stable toward light and heat, but stability to sulphur dioxide is poor. In powdered form, this pigment can be used for coloring various instant foodstuffs and in alkaline solutions and in ammonia for coloring different foodstuffs, including baked products, yogurts, soups, desserts, confectionery, and syrups.

SOME OTHER NATURAL PIGMENTS

Riboflavin and Riboflavin 5′ Phosphate (E 101)

Riboflavin—Vitamin B_2 (Formula 8.15)—is a yellow pigment present in many products of plant and animal origin. Milk and yeast are the best sources of riboflavin. It is an orange-yellow crystalline powder, intensively bitter tasting, which is very slightly soluble in water and ethanol, affording a bright greenish-yellow fluorescent solution. Riboflavin is stable under acid conditions but unstable in alkaline solution and when exposed to light. Reduction produces a colorless leuco form, but color is regenerated again in the contact with air.

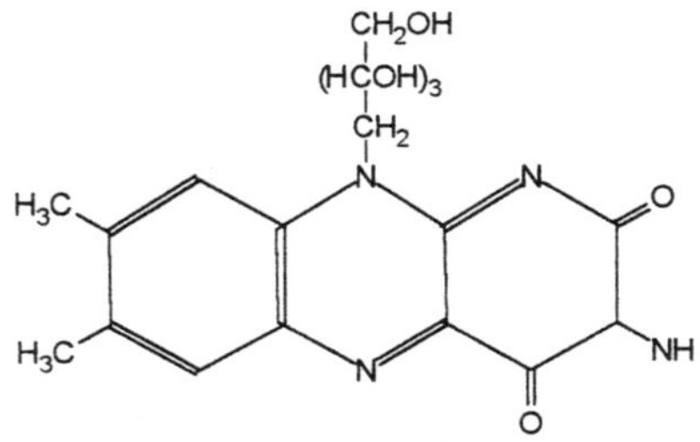

$$CH_2OH$$
$$(HCOH)_3$$
$$CH_2$$

(8.15)

Riboflavin

Riboflavin-5′-phosphate sodium salt is much more soluble in water than the unesterified riboflavin and is not so intensely bitter. It is one of the physiologically active forms of vitamin B_2. It is rather more unstable to light than riboflavin.

Turmeric, Curcumin (E 100)

Turmeric, or curcumin, is the fluorescent yellow-colored extract from the rhizome of various species of curcuma plant, with *Curcuma longa* L. giving the best coloring matter. The main pigment of curcuma is curcumin (Formula 8.16).

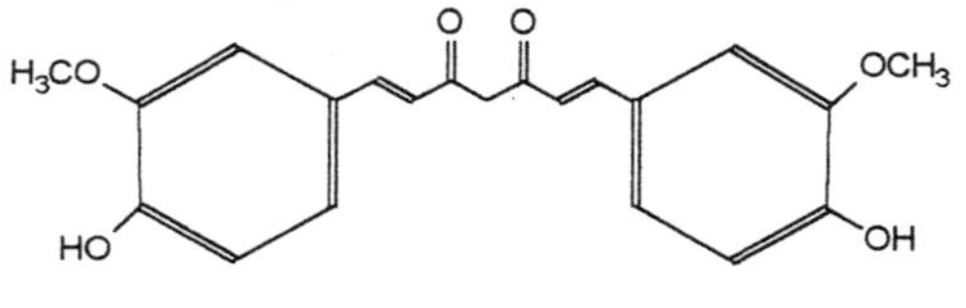

(8.16)

Curcumin

Turmeric oleoresin is insoluble in water but soluble in alkalis, alcohols, and glacial acetic acid. This pigment has a strong characteristic odor and sharp taste and is utilized for both its taste and color properties as an additive to canned products, soups, mustards, and other products.

Caramel (E 150)

Caramel is the amorphous dark brown coloring material formed by heating saccharides in the presence of selected accelerators. It consists of a mixture of volatile and nonvolatile low-molecular-weight compounds and high-molecular-weight compounds. The composition and coloring power of caramel depends on the type of raw material and process used. Both Maillard-type reaction and caramelizing reactions are involved, and the commercial products are extremely complex in composition. Detailed information on caramel is given in Chapter 4.

SYNTHETIC ORGANIC COLORS

The synthetic food colorants, according to their chemical structure, belong to mono-, dis-, and trisazo; triarylmethane; xanthene; quinoline, and indigoid. They can also be divided into water-soluble, oil-soluble, insoluble (pigment), and surface marking colors.

Water solubility is conferred on many dyes by introducing to the molecule at least one salt-forming group. The most commonly used is the sulphonic group, but carboxylic acid residues can also be used. These dyes are usually isolated as sodium salts. They have a colored anion and are known as anionic dyes. The other dyes containing basic groups, such as $-NH_2$, $-NH-CH_3$, or $-N(CH_3)$, form water-soluble salts with acids. These are the cationic dyes, and the colored ion is positively charged. If both acidic and basic groups are present, an internal salt is formed.

Oil soluble or solvent-soluble colors lack salt-forming groups. Pigments are the colors having no affinity for most substrates. They are thus generally insoluble in water, fats, and solvents, so they color by dispersion in the food medium. Precipitation of water-soluble colors, with aluminum, calcium, or magnesium salts (generally with aluminum) forms water-insoluble pigments called lakes. Lakes may be prepared from all classes of water-soluble food colors, and they are the most important group of food color pigments.

The stability of synthetic food colors towards the condition prevailing in food processing depends upon product composition, temperature, and time of exposition and others. Generally, they are resistant to boiling and baking conditions, but light has a destructive effect on all of the colors. The use of synthetic colors is the most reliable and economical method of coloring those products that have little or no natural color present, such as dessert powders, table jellies, and sugar confectionery.

The synthetic organic dyes are superior to the natural colorants in consistency of strength and range and brilliance of shade, as well as in stability, ease of application, and cost-effectiveness. However, the manner in which synthetic colorants were employed, from a safety viewpoint, left much to be desired; therefore, regulations were introduced to control the use of these added food colorants.

REFERENCES

Brouillard, R., 1981. "Origin of Exceptional Color Stability of the Zebrina Anthocyanin," *Phytochemistry,* 20, 143–147.

Brouillard, R., 1982. "Chemical Structure of Anthocyanins," in *Anthocyanins as Food Colors,* Ed. Markakis, P., New York: Academic Press, pp. 1–38.

Brouillard, R., Wigand, M. C., Dangles, O., Cheminat, A., 1991. "pH and Solvent Effects on the Copigmentation Reaction of Malvin by Polyphenols, Purine and Pyrimidine Derivatives," *J. Chem. Soc. Perkin Trans.,* 2(8), 1235–1241.

Combs, G. F., Jr., 1992. *The Vitamins—Fundamental Aspects in Nutrition and Health,* San Diego: Academic Press Inc., p. 121.

Coulson, J., 1980. "Miscellaneous Naturally Occurring Colouring Materials for Foodstuffs," in *Developments in Food Colours, Vol. 1,* Ed. Walford, J., London: Applied Science Publishers LTD, pp. 189–218.

Czapski, J. 1985. "The Effect of Heating Conditions on Losses and Regeneration of Betacyanins," *Z. Lebensm. Unters. Forsch.,* 180, 21–25.

Goto, T., Kondo, T., Tamura, H., Iagawa, H., Iino, A., Takeda, K., 1982. "Structure of Gentiodelphin, an Acylated Anthocyanin Isolated from *Gentiana makinoi,* that is Stable in Diluted Aqueous Solution," *Tetrahedron Lett.,* 23, 3695–3698.

Humphrey, A. M., 1980. "Chlorophyll." *Food. Chem.,* 5, 57–67.

Klaui, H., Bauernfeind, J. C., 1981. "Carotenoids as Food Colors," in *Carotenoids as Colorants and Vitamin A Precursors,* New York: Academic Press, pp. 47–318.

Macheix, J. J., Fleuriet, A., Billot, J., 1990. "Importance and Roles of Phenolic Compounds in Fruits," in *Fruits Phenolics,* Boca Raton, Florida: CRC Press, pp. 239–278.

Mazza, G., Miniati, E., 1993. *Anthocyanins in Fruits, Vegetables and Grains,* Boca Raton: CRC Press, pp. 1–23.

CHAPTER 9

Flavor Compounds

CHUNG-MAY WU
BONNIE SUN PAN

SOURCES OF FLAVORS IN FOODS

FLAVORS FORMED NATURALLY IN PLANTS

Spices and Herbs

SINCE antiquity, spices, herbs, and condiments have been considered virtually indispensable in the culinary arts. They have been used to flavor foods and beverages the world over. Spices can be grouped according to the parts of plant used: leaves (bay, laurel), fruits (allspice, anise, capsicum, caraway, coriander, cumin, dill, fennel, paprika, pepper), arils (mace), stigmas (saffron), flowers (safflower), seeds (cardamom, celery seed, fenugreek, mustard, poppy, sesame), barks (cassia, cinnamon), buds (clove, scallion), roots (horseradish, lovage), and rhizomes (ginger, turmeric). Most of the spices and herbs contain volatile oils, called essential oils, which are responsible for the characteristic aroma of spices. Some spices (capsicum, ginger, mustard, pepper, horseradish) are pungent, while paprika, saffron, safflower, and turmeric are valued for their colors. Many spices have some antioxidant activities. Rosemary and sage are particularly pronounced in antioxidant effects. Cloves, cinnamon, mustard seed, and garlic contain antimicrobial activities. Some spices have physiological and medicinal effects. Spiced foods contain substances that affect the salivary glands (Pruthi, 1980).

Fruits

Citrus fruits contain peel oil, the essence from which oil is obtained

211

during concentration of the juice process. Citrus oils are characterized by a high percentage of terpene hydrocarbons (limonene, $C_{10}H_{16}$), which contribute little to aroma. The unique characteristics of limonene are its relative insolubility in dilute alcohol and its susceptibility to oxidation, causing off-flavor production. If the monoterpenes are removed, the resulting oil is called terpene-free or terpeneless oil. Aldehydes, esters, and alcohols are the main contributors to the aromas of citrus oil. These compounds are relatively polar and soluble in water; therefore, they are satisfactory for applications in food and beverage.

Fruits other than citrus contain much less volatile aromatic compounds and cannot form essential oil in distillates. However, their juices, juice concentrates, extracts of dehydrated fruits, and distillates (essence) can be used as flavorants directly added to foods.

Vegetables

Vegetables contain some flavor compounds, the concentrations of which are mostly too low to obtain essential oils. In the tissues of some vegetables, volatile compounds are enzymatically produced when they are disrupted. Vegetables have the function of flavoring only after their cells are disrupted or after being oil-fried. Vegetable flavors are classified in the category of savory flavor, while fruit flavors are classified as sweet flavors.

In addition to the three natural flavor categories described above, tea, coffee bean, cocoa bean, flowers (i.e., rose, jasmine), peppermint, and balsam are natural products with flavoring properties (Arctander, 1960; Furia and Bellanca, 1975).

FLAVORS PRODUCED BY MICROBES OR ENZYMES

Fermentation has been known and commercially exploited for centuries. Products like spirits, liqueurs, wines, beer, and other alcoholic beverages; vinegar; cheese and yogurt; miso, soy sauce, and fermented bean curd; hams and sausages; fish seasonings; curing of vanilla beans, tea, and cocoa; pickles and sauerkraut; dough, bread, and other bakery products have special flavor notes that can also be used as seasonings.

Biotechnology for the production of flavoring materials has been developed in the past decade. The technology relating to the production of flavors include cell and tissue culture; microbial fermentation; and bioconversion of substrates using whole microbial cells, plant cells, or enzymes (Harlander, 1994).

FLAVOR PRODUCED BY HEATING OR COOKING

The flavor of foods such as wheat, peanut, and sesame, after being cooked, are quite different from those of the raw materials. Flavor formation from flavor precursors in the processed foods are primarily via Maillard reaction, caramelization, thermal degradation, and lipid-Maillard interactions.

FLAVORS FROM FLAVORANTS ADDED

Flavorings play essential roles in the production of a wide range of food products versatile in aroma to allow consumer choices and to meet consumer needs. In this regard, flavor manufacturers require expertise in flavor formulations, research, and technical services, while flavor users need fundamental knowledge of flavor applications.

MOLECULAR STRUCTURE AND ODOR OF FLAVOR COMPOUNDS

VOLATILITY AND INTENSITY OF AROMA COMPOUNDS

The flavor of food is largely perceived as a result of the release of odorous compounds, usually present in trace amounts in foods, into the air in the mouth and, thence, to the olfactory epithelium in the nose. The volatility is caused by the evaporation or the more or less rapid sublimation of an odoriferous substance. It is proportional to the vapor pressure of the substance and inversely proportional to its molecular weight. Therefore, an aroma compound must be volatile. Other characteristics relate to aroma compounds; among them flavor intensity is the most important one. Threshold is used most extensively for quantification of flavor intensity. Psychophysically, a threshold can be defined as the minimum concentration of a stimulus that can be detected (absolute threshold), discriminated (just-noticeable difference), or recognized (recognition threshold). In general, the detection thresholds are lower than the recognition thresholds, if the difficulty in measuring both are comparable (Pangborn, 1981).

The relationship between the molecular structure of an aroma compound and its threshold is still unclear. Volatility of a compound may not relate to its threshold. For example, the threshold of ethanol (boiling point is 78°C) is much higher than octanol (boiling point is 195°C) or other homologous alcohol. Ethanol has high volatility but low odor intensity. It is often used as a solvent in compounded flavors.

In general, sulphur- or nitrogen-containing compounds and heterocyclic compounds have very low threshold values.

Odor unit is defined as the ratio of concentration and threshold. This unit quantifies the contribution of a specific component or a fraction to the total odor of a mixture; however, it says nothing about the odor quality of the final mixture, nor does it imply anything about the relationship between the stimulus concentration and the intensity of sensation above the threshold (Teranishi et al., 1981).

FLAVOR COMPOUNDS AND THEIR ODORS

The relationship between the molecular structure of a chemical compound and its odor has been the subject of much research and conjecture. It is still not possible to predict the aromatic profile from the structure of a given chemical, nor is it possible to assume changes in flavor profile based on molecular structure modification. Even stereo-isomers may differ in odor both qualitatively and quantitatively. Nevertheless, the relationship between structure and odor can be summarized as follows.

Small molecules — such as ethanol, propanol, butanol among alcohols; acetaldehyde, propionaldehyde among aldehydes; and acetic acid, propionic acid, butyric acid among acids — are highly volatile and exhibit pungent ethereal, diffusive, harsh, or chemical odor characteristics. Only in extreme dilution of these compounds will desirable odor become perceptible. Bigger molecules, from 5°C to 10°C of alcohols, aldehydes, and acids, are mild and desirable. Molecules containing more than 10°C of alcohols, aldehydes, and acids reduce their volatility and odor intensities with increases in molecular size.

Compounds containing functional groups such as $-OH$, $-CHO$, $-CO$, and $-COOH$, play important roles in exporting the odors of a compound. Acid is sour, aldehyde yields fresh note, and ester is fruity. However, elongated alkyl group enhances the fatty or oily note. Ketones having two alkyl groups attached to a carbonyl group give more fatty aroma than the corresponding aldehydes.

Ester has a fruity note. When the initial alkyl group in alcohol or acid or both is relatively large in molecular size or with its own characteristic note, the resulting ester maintains this note in addition to the fruity note. Examples are citronellyl acetate, having the fresh-rosy-fruity odor, inherits the rosy note from citronellol; bornyl acetate, having a sweet herbaceous-piney odor with a balsamic undertone, maintains the odor of borneol.

The boiling point of ethyl acetate is 77°C, and its molecular weight is 88. Those of its reactant ethyl alcohol are 78°C and 46, while those of the acetic acid are 118°C and 60, respectively. Ester has a higher molecular weight but a lower boiling point, in comparison to the precursor alcohol and acid. Many esters such as ethyl acetate have fresh note. Aldehyde has a relatively low boiling point. For example, acetaldehyde has a molecular weight of 44, but its boiling point is 21°C. Therefore, aldehydes are often used due to their fresh note. Decanal contributes to the fresh note in orange aroma.

Lactones are cyclic compounds with an ester functional group. They have the characteristic ester notes: fruity, oily, and sweet. However they are cyclic compounds and have a relatively high boiling point. γ-Undecalactone, with a peach-like aroma, has a boiling point of 297°C.

Heterocyclic compounds, generally have very low thresholds. Thiazoles, thiolanes, thiophenes, furans, pyrazines, and pyridines are normally present in larger numbers and a higher concentration in cooked, fermented, or processed seafoods or meat products than in fresh ones (Pan and Kuo, 1994).

Essential oil, oleoresin, or other natural flavoring raw materials have many valuable trace components, which play important roles in aroma. These components are not commercially available now because of their complexity and low threshold. It is not feasible to undergo a complicated manufacturing process for the very little amount needed. The only source available is the natural product.

CHANGES IN FLAVOR DURING FOOD STORAGE AND PROCESSING

CHANGES DUE TO NATURE OF FLAVOR COMPOUNDS

A volatile compound evaporates continuously, even at room temperature. Higher temperature accelerates the evaporation. Some food ingredients such as lipids and proteins may trap flavor compounds to some extent and reduce their volatility. Different flavor compounds have different volatility. An aged food may lose, not only its total flavor, but also change its proportions of the flavor components, resulting in a changed odor.

Many flavor compounds contain double bonds or aldehyde groups, which are susceptible to oxidation, cleavage, or polymerization. Alcohols can be oxidized to the corresponding aldehyde and then acid. Alcohol and acid can react and dehydrate to form an ester. Ester can be hydrolyzed to alcohol and acid at neutral or alkaline pH. Aldehyde and alcohol can be dehydrated by catalysis to form hemiacetal, and the reverse reaction can occur in acidic condition or in water.

CHANGES DUE TO CONTINUING AROMA BIOGENESIS

The amount of secondary metabolites such as aroma compounds produced by a plant during its life cycle is a balance between formation and elimination. The two opposing functions are directly controlled by two main groups of factors. The intrinsic factors are comprised of all internal, or hereditary, properties (e.g., genotype and ontogeny), while extrinsic factors are comprised of all external, or environmental, properties (e.g., pressure, wind, light, temperature, soil, water, nutrients). Therefore, a plant material such as citrus fruits (Nagy and Shaw, 1990) or the essential oils (Lawrence, 1986) may have quite a different flavor quality due to the culture conditions and maturities. The typical flavor of climateric fruit such as banana, peach, pear, and cherry is not present during early fruit formation, but develops entirely during a rather short period of ripening, during which time minute quantities of lipid, carbohydrate, protein, and amino acids are enzymatically converted to volatile flavors (Reineccius, 1994c). During postharvest handling, the plant continues the biogenesis of aroma.

CHANGES DUE TO TISSUE DISRUPTION OR ENZYME REACTIONS

Introduction

Some food flavors are not present in the intact plant tissues, but are formed by enzymatic processes when the plants are cut or crushed. Under these circumstances, the cells are ruptured, and the flavor precursors are released and exposed to enzymes. Unique examples of this kind of flavor formation are described below.

Alliums

Garlic, onion, shallot, green onion, and chive belong to the genus *Allium*. Members of this genus contain volatile sulphur compounds, including thiols, sulfides, disulfides, trisulfides, and thiosulfinates (Block and Calvey, 1994; Yamaguchi and Wu, 1975; Yu and Wu, 1981).

The enzymatic flavor formation reaction of the genus *Allium* can be generalized as follows:

$$2RSOCH_2CHNH_2COOH \rightarrow RSSOR + 2NH_3 + CH_3COCOOH$$

where R = methyl, propyl, 1-propenyl, or allyl. The 1-propenyl compound has been identified as the lacrimator in onions. Allicin, the active odor principle of fresh garlic, is diallyl thiosulfinate. In common with all thiosulfinates, allicin readily forms diallyl disulfide and diallyl trisulfide at room temperature. Addition of soybean oil in the process of garlic disruption can slow down the conversion of allicin (Kim et al., 1995).

Brassicas

The *Brassicas* of importance as foods include turnips, rutabagas, mustards, and the cole crops—cabbage, broccoli, cauliflower, and brussels sprouts. The production of isothiocyanates in *Brassicas* is via an enzymatic reaction on specific glycosides. Some of the isothiocyanates, especially allylthiocyanate, are highly pungent and are mainly responsible for the odor of brown mustard, horseradish, cabbage, and other crucifers. Any process that destroys or inactivates enzymes in these plants will cause decreases in aroma production, resulting in a less distinctive flavor. This is usually the case when *Brassica* foodstuffs are commercially preserved.

Mushrooms

1-Octen-3-ol occurs in many mushroom species. It contributes significantly to the flavor of edible mushrooms such as *Agaricus campestris* (Tressl et al., 1982), and *Agaricus bisporus* (Wurzenberger and Grosch, 1982; Chen and Wu, 1984). 1-Octen-3-ol is formed enzymatically from linoleic acid, which was shown to be the major fatty acid in *A. bisporus*. Enzymes involved in the pathway of formation of 1-octen-3-ol include lipoxygenase, hydroperoxide cleavage enzyme, and alcohol oxidoreductase (Tressl et al., 1982).

Shiitake (*Lentinus edodes*) is one kind of edible mushroom highly prized in China and Japan. Due to the difficulties of postharvest storage, the mushroom has been traditionally preserved in dried form. The differences between the fresh shiitake and the dried shiitake lie in the contents of eight-carbon compounds (i.e., 3-octanone, 1-octen-3-ol, 3-octanol, *n*-octanol, and *cis*-2-octen-1-ol) and sulphurous compounds (i.e., dimethyl disulfide and dimethyl trisulfide). Fresh shiitake contains more eight-carbon compounds than dried shiitake does, while dried shiitake contains more sulphurous compounds than the fresh (Chang et al., 1991). The formation of volatile shiitake is affected greatly by the pH during blending. 1-Octen-3-ol and 2-octen-1-ol are predominantly formed around pH 5.0–5.5, while the formation of predominantly sulphurous compounds such as dimethyl disulfide and dimethyl trisulfide is around pH 7.0. Two enzymatic systems are probably responsible for the formation of eight-carbon and sulphurous compounds. Therefore, an enzymatic reaction is likely to occur in dried shiitake, yielding more volatile sulphur compounds than the fresh shiitake (Chen et al., 1984).

Formation of Green-Grassy Notes in Disrupted Tissues

Six-carbon compounds such as hexanal, 3Z-, and 2E-hexenal at high concentrations were detected in ruptured tissue of apples, grapes, and tomatoes (Schreier and Lorenz, 1981). These compounds occur only in trace amounts or none in the intact plant cells. Aliphatic C-6 components, which contribute to the green note of fruits, are formed from unsaturated C-18 fatty acids by enzymatic activity after cellular disruption. Lipoxygenase is involved in the reaction (Galliard and Matthew, 1977). (Z)-3-Hexen-1-ol, (E)-2-hexenal, hexanol, (E)-2-hexen-1-ol, and hexanal are formed in bell peppers (*Capsicum annuun* Var. *grossum,* Sendt) after tissue disruption (Buttery and Ling, 1992).

Glycosides as Flavor Precursors

There are two forms of monoterpene derivatives in grapes: free and glycosidic conjugates. The free form consists of compounds with interesting flavor properties such as geraniol, nerol, linalool, linalool oxides, α-terpineol, cirronellol, hotrienol, and flavorless polyhydroxylated compounds (polyols), which under mild acid hydrolysis conditions can yield odorous volatiles. The flavorless glycoside forms consist of β-D- glucopyranosides and diglycosides; 6-O-α-L-arabinofuranosyl-β-D-glucopyranosides, 6-O-α-L-rhamnopyranosyl-β-D-glucopyranosides (rutinosides), and 6-O-β-D-apiofuranosyl-β-D-glucopyranosides of predominantly geraniol, nerol, and linalool, together with monoterpenes, are present at a higher oxidation state than the free forms. As most of these compounds have interesting sensory properties, their glycosides make up a potential aroma reserve more abundant than their free counterparts. The glycosidically bound volatiles can be released by either acid or enzyme hydrolysis (Wu and Liou, 1986). β-Glucosidase, being the most abundant glycosidase, is present with α-arabinosidase and α-rhamnosidase in grape berries of various cultivars. Enzyme treatment of juice or wine increases the concentrations of volatile monoterpene flavorants. Prolonged aging of wine or its exposure to elevated temperatures increases the concentration of free volatile monoterpenes through hydrolysis of glycosidic precursors in wine (Gunata et al., 1992; Strauss et al., 1986).

Other plants such as papaya (Schreier and Winterhalter, 1986), nectarine (Takeoka et al., 1992), and tea leaves (Kobayashi et al., 1992) also contain glycosides as precursors of flavors.

CHANGES DUE TO PROCESSING

Maillard Reaction

The Maillard reaction plays an important role in flavor development. Products of Maillard reaction are aldehydes, acids, sulphur compounds (e.g., hydrogen sulfide, methanethiol), nitrogen compounds (e.g., ammonia, amines), and heterocyclic compounds such as furans, pyrazines, pyrroles, pyridines, imidazoles, oxazoles, thiazoles, thiophenes, di- and trithiolanes, di- and trithianes, furanthiols, etc. Higher temperature results in production of more heterocyclic compounds among which many have roasty, toasty, or caramel-like aroma.

Sugar, ascorbic acid, amino acids, thiamine (de Ross, 1992; Ames and

Hincelin, 1992, Güntert et al., 1992, 1994), and peptides (Ho et al., 1992; Izzo et al., 1992; de Kok and Rosing, 1994) are potential reactants of the Maillard reaction. They are present in most foods, so the Maillard reaction occurs commonly when these foods are cooked. The conditions of cooking determine the aroma of the cooked foods. For example, the major volatiles identified from water-boiled duck meat were the common degradation products of fatty acids, while roasted duck meat contains not only the volatiles found in raw duck meat, but also pyrazines, pyridines, thiazoles, etc. (Wu and Liou, 1992), which are Maillard reaction products. The wax gourd (*Benincasa hispida,* Cogn), also known as winter melon or gourd melon, a vegetable, is used to produce beverages, candy, or jam, which are popular in Taiwan. The flesh of the gourd melon is white in color. The major volatile compounds of fresh gourds are (E)-2-hexenal, *n*-hexanal, and *n*-hexyl formate, while the 2,5-dimethylpyrazine, 2,6-dimethylpyrazine, 2,3,5-trimethylpyrazine, 2-methyl pyrazine, and 2-ethyl-5-methyl pyrazine are the major volatile compounds of the wax gourd beverage, which is brown in color. The beverage is prepared by cooking sliced wax gourd and sugar at alkaline pH for about 3–4 hours or even longer and then diluting with water and serving as a nonalcoholic beverage. The pyrazine compounds not present in the fresh wax gourd are likely formed from the sugar added and the endogenous amino acids during processing of the beverage (Wu et al., 1987). This is an example to show changes in flavor of foods during processing while the Maillard reaction plays an important role of the changes.

Hundreds of patents have been granted worldwide for processes and reaction products based on Maillard technology applied to meat and savory flavors (Buckholz, 1988; Mottram and Salter, 1988; Ouweland et al., 1988).

Maillard reaction may produce mutagenic components, pigments, and antioxidants, which are discussed in other sections of this book.

Lipid Oxidation

The oxidation products of lipids are volatile aldehydes, acids, etc. Therefore, lipids are one of the major sources of flavors in foods. For example, much of the desirable flavors of vegetables such as tomatoes, cucumbers, mushrooms, and peas (Ho and Chen, 1994); fresh fish (Hsieh and Kinsella, 1989); and cooked shrimp (Kuo and Pan, 1991; Kuo et al., 1994), as well as many deep-fat fried foods such as French-fried potatoes (Salinas et al., 1994) and fried chicken (Shi and Ho, 1994) are contributed to by lipid oxidation. Lipoxygenase-catalyzed lipid oxidation produces secondary derivatives, i.e., tetradecatrienone, which is a keynote compound of shrimp (Kuo and Pan, 1991).

The major difference between the flavor of chicken broth and that of beef broth is the abundance of 2,4-decadienal and γ-dodecalactone in chicken broth (Shi and Ho, 1994). Both compounds are well known lipid oxidation products. A total of 193 compounds has been reported in the flavor of chicken. Forty-one of them are lipid-derived aldehydes.

The core flavor of mashed potatoes consists of naturally occurring and thermally generated compounds. These compounds arise mainly from the oxidation of fatty acids, especially highly unsaturated fatty acids, and from the degradation and interaction of sugar-amino acids. The extent to which these reactions affect the flavor of the final product depends on the age of the raw materials, storage conditions and processing techniques. For example, both lipid oxidation and nonenzymatic browning reactions increase with the age of the raw potato (Salinas et al., 1994).

Garlic develops its aroma from enzymatic reactions, as described before. When garlic slices are deep-oil fried, microwave heated, or oven baked, the aroma changes (Yu et al., 1993) and contributes to a different kind of garlic flavor to foods. A novel S-compound was identified from the interaction of garlic and heated edible oil (Hsu et al., 1993). Also, alliin and deoxyalliin, two important flavor precursors of garlic, can react with 2,4-decadienal, which is one of the major oxidation products of fatty acids, to form aroma (Yu et al., 1994).

Deep-fat frying is a universal cooking method. Stir frying is common in some cuisines, especially in Chinese cooking. Changes of volatile compounds in oils after deep-fat frying or stir frying and subsequent storage were studied (Wu and Chen, 1992). Soybean oil (900 g) was heated by deep frying at 200°C for 1 hr with the addition of 0, 50, 100, 150, and 200 ml water, respectively, and then stored at 55°C for 26 weeks. All samples contained aldehydes as major volatiles. During heating and storage, total volatiles increased 260–1100 fold. However, aldehyde content decreased from 62–87% to 47–67%, while volatile acid content increased from 1–6% to 12–33%. Hexanoic acid increased to 26–350 ppm in the oils after storage. Hexanoic acid has a heavy, acrid-acid, fatty, rancid odor, often described as "sweat-like," which is responsible for the "rancid" note. Water addition to the oils deep fried tended to retard the formation of volatile compounds.

Freshly stir-fried Chinese food has a much better flavor quality than after it is aged. The main change in volatile constituents of stir fried bell peppers during aging is the production of volatile carbonyl compounds from autoxidative breakdown of unsaturated fatty acids (Wu et al., 1986).

Generally, the undesirable flavor qualities of food are associated more closely with lipids than with proteins and carbohydrates. Lipids are responsible for the rancidity of lipid-containing foods. The term *warmed-over flavor* (WOF) is used to describe the rapid development of oxidized odor in cooked meat upon subsequent holding. The rancid or stale flavor becomes readily apparent within 48 hr, in contrast to the more slowly developing rancidity that becomes evident only after frozen storage for a period of months. Although WOF was first recognized as occurring in cooked meat, hence the name WOF, it also develops in raw meat that is ground and exposed to air. Overheating of meat protects it against WOF by producing Maillard reaction products possessing antioxidant activity (Pearson and Gray, 1983).

Flavor chemistry of lipid foods has been reviewed and compiled elsewhere in recent years (Ho and Chen, 1994; Min and Smouse, 1989).

The Interaction of Lipids in the Maillard Reaction

The Maillard reaction and the oxidation of lipids are two of the most important reactions for the formation of aromas in cooked foods. Interactions between lipid oxidation and the Maillard reaction have received less attention, despite the fact that lipids, sugars, and amino acids exist in close proximity in most foods. Lipids, upon exposure to heat and oxygen, are known to decompose into secondary products, including alcohols, aldehydes, ketones, carboxylic acids, and hydrocarbons. Aldehydes and ketones produce heterocyclic flavor compounds reacting with amines and amino acids, via Maillard-type reactions in cooked foods (Shibamoto and Yeo, 1992). Lipid degradation products such as 2,4-decadienal and hexanal can interact with Maillard reaction intermediates to form long-chain alkylpyrazines, as well as other heterocyclic compounds (Farmer and Whitefield, 1993).

Extrusion

Extrusion-cooking is a process whereby foodstuffs of low-moisture content (10–30%) are submitted to the action of heat, pressure, and mechanical shearing during a short time (20 sec to 2 min). Extrusion can have a significant effect on flavor and aroma profiles of food products manufactured through this process. Depending on raw material composition, flavor development during the process may be an important consideration in product quality. Certain mechanisms such as nonenzymatic browning and lipid oxidation are considered to have significant implications in the flavor characteristics of food products. Oxidation and volatility of flavor compounds are important factors to be taken into consideration during heating and extrusion cooking at different temperatures and moisture contents. Lipid oxidation products are the major compounds of aroma generation in the extrudates prepared from wheat flour at high moisture content and low die temperature. By lowering moisture content, lipid degradation compounds decreased, and the Maillard reaction products dominated the flavor profile. The lipid oxidation products significantly increased in the sample extruded at low moisture content and high die temperature during storage (Villota and Hawkes, 1988).

Retention of aroma compounds during extrusion-cooking of different formulations — starch, starch-caseinate, biscuit mix — was studied (Sadafian and Crouzet, 1987). Several aroma compounds — limonene, p-cymene, linalool, geraniol, terpenyl acetate, and β-ionone — were added in different ways: water emulsion, oil solution, capsules, or inclusion complexes in β-cyclodextrin. During the extrusion process, major loss of free volatiles reaches to more than 90%. It is controlled by water stripping during the expansion phase of the extrudate. Flavor retention is increased through encapsulation of volatile compounds in natural or artificial walls or by use of inclusion complexes in β-ionone.

Concentration and Other Processes

Some foods have special treatment in processing, which may affect the composition of volatile components. As an example, in hybrid passion fruit, the presence of about 1-2% starch makes heat processing, i.e., pasteurization and concentration, impossible or impractical, unless the starch is removed before processing. However, the step of removing starch and concentration causes loss of volatile compounds (Kuo et al., 1985).

CHANGES DUE TO STORAGE OF FOOD PRODUCTS

The ability to extend the period of availability of foods and food combinations in preserved forms has improved man's health, added variety to his diet, and reduced the time for food preparation. The basic preserving processes are canning, freezing, dehydration, salting, pickling, and freeze-drying. Food preservation is designed to prevent undesirable changes in food and food products. However, flavor changes in food products during storage occur continuously for processed foods although the deterioration of flavor quality is not significant in most cases.

Nonenzymatic reactions that occur during processing and storage of food products are detrimental if foods contain reducing compounds or if these compounds are produced during storage as a result of oxidation, acid hydrolysis, enzymatic reactions, or physico-chemical changes. Water mediates the nonenzymatic browning reaction by controlling the liquid phase viscosity; by dissolution, concentration, and dilution of reactants; and by effects on the reaction pathways due to activation energy limitations in dehydrated foods (Saltmarch et al., 1981).

Citrus juice can have the problem of off-flavor formation during processing and storage. Changes in volatile components in aseptically packaged orange juice during storage at 21 °C and 26°C were monitored. Quantities of several desirable flavor components decreased during storage, while amounts of two undesirable components, α-terpineol and furfural, increased progressively with length of storage (Moshonas and Shaw, 1987).

The ultra high temperature (UHT) processing of milk owes its commercial success to the observation that the rate of destruction of microorganisms increases more rapidly with temperature than do the rates of the accompanying color and flavor changes. At very high processing temperatures, high sterility may be achieved with minimal adverse nutritional and chemical effects. However, UHT milk darkens in color during storage. This effect is noticeable after a few months of storage at 20°C. It becomes more pronounced at higher temperature and longer storage time. The milk also deteriorates in taste. The ϵ-amino group of lysine in milk proteins may react extensively with lactose by the Maillard reaction before the milk develops marked off-flavor, discoloration, or instability (Moller, 1981). Spray dried whey also has the problem of browning via Maillard reaction.

USE OF FLAVORS IN FOOD INDUSTRY

INTRODUCTION

Taste, aroma, texture, and visual appearance play very important parts in the appeal of all prepared foods produced domestically or commercially. Food flavorings are compounded from natural and/or synthetic aromatic substances. The compounded flavor may or may not be found in nature. Reasons for using flavors in foods are the following (Giese, 1994):

- Flavors can be used to create a totally new taste. This does not happen very often, but some new flavors have been enormously successful such as those used by Coca-Cola® or Pepsi Cola®.
- Flavoring ingredients may be used to enhance, extend, round cut, or increase the potency of flavors already present.
- Processing operations such as heating may cause a loss of flavor, while some flavors already present may need supplementation or strengthening.
- Flavor ingredients can simulate other more expensive flavors or replace unavailable flavors.
- Flavors may be used to mask less desirable flavors. This function does not imply that flavors are being used to hide spoilage but, rather, are used to cover harsh or undesired tastes naturally present in some processed foods.

The main activities of flavor industries are collection or production of flavoring materials, flavor manufacturing, flavor application studies, and technical services.

COLLECTION OR PRODUCTION OF FLAVORING MATERIALS

Natural Flavor Materials

The sources, names, characteristics, and major flavor components of natural flavoring materials such as spice, herb, etc. have been summarized in several books (Arctander, 1960; Furia and Bellanka, 1975; Reineccius, 1994b). A large portion of constituents in natural flavor materials is not flavor compounds. These nonflavor compounds have to be removed to produce the concentrated flavorants. There are two major methods to reach this purpose as follows.

Distillation

The essential oils are the distilled fraction of aromatic plants. Most often, they are steam distilled. These oils, which are primarily responsible for the characteristic aroma of the plant material, are generally complex mixtures of organic compounds. During the concentration of citrus juices, a layer of essential oil is formed in the condensate. This oil is called essence oil. Terpenes can be removed from both essential oil and essence oil to obtain folded oil. Some fruit or vegetable distillates containing flavor compounds are called essence. Although no essential oil layer is obtained from vegetable distillate, it is used as a flavoring raw material. Citrus peels, rich in essential oils, are expressed to get essential oils, which are called cold pressed oils.

Solvent Extraction

Essential oils do not contain hydrophilic flavoring components, antioxidants, or pigments. The nonvolatile flavoring constituents of aroma plants are recoverable by extraction. The selection of solvent is limited, depending on its toxicity, and whether or not it remains in the final product. Two kinds of solvents are used: 1) a polar solvent such as ethyl alcohol; for example, vanillin is ethyl alcohol-soluble, therefore, ethyl alcohol is used as the solvent to prepare vanilla bean extract; 2) a nonpolar solvent such as petroleum ether. Most aroma compounds are oil-soluble. Therefore, petroleum ether is used as the solvent to extract plant aroma. The extract, after removal of the solvent, is called concrete. Concrete may contain wax and fatty acids in large proportion and is further purified by ethyl alcohol extraction. The product is called absolute. The nonvolatile flavoring constituents of herbs and spices are recoverable by extraction. In practice, a solvent is chosen that dissolves both the essential oil and the nonvolatiles present. The resulting solvent-free product is known as oleoresin. A disadvantage of the oleoresins is that they are very viscous and thick, making them difficult to handle or to mix in processing operations. Several products have been developed into extractives, being convenient to use and avoiding handling problems. Extractives can be dispersed in salt, dextrose, or other carriers to create dry-soluble spices. They may also be dispersed in fats to make fat-based soluble spices. Emulsification of extracts with starches and gums and spray drying produces encapsulated spices. Solubilization of extracts with glycerol, isopropyl alcohol, and propylene glycol produces liquid-soluble spices (Giese, 1994b).

Organic Chemicals Used in Flavorings

Organic chemicals being used in flavorings include hydrocarbons, such as limonene, pinene, ocimene, α-phellandrene, β-caryophyllene; alcohols, such as hexanol, *cis*-3-hexen-1-ol, geraniol, citronellol, eugenol, 1-menthol; aldehydes, such as acetaldehyde, hexanal, 2,4-decadienal, citral, vanillin; ketones, such as diacetyl, ionone, nootkatone; acids, such as acetic acid, butyric acid, pyroligenious acid; esters, such as ethyl acetate, linalyl acetate, ethyl phenyl acetate, methyl dihydrojasmonate; lactones, such as γ-nonalactone, δ-decalactone, γ-undecalactone; hemiacetals, such as acetaldehyde diethylacetal, citral dimethyl acetal; ethers, such as diphenyl oxide, rose oxide; nitrogen-containing compounds, such as trimethylamine; sulphur-containing compounds, such as dimethylsulfide, thiolactic acid, allyl disulfide; and heterocyclic compounds, such as furans, pyrazines, pyridines, thiazoles. The names, chemical structures, physical and organoleptic properties, and uses have been summarized in several books (Arctander, 1969; Furia and Bellanca, 1975; Reineccius, 1994b). Many organic chemicals being used in flavorings are produced by synthetic method and are commercially available. More and more natural compounds are used in flavorings due to the fact that the compounded flavors can be classified as natural flavorings. They are produced or prepared by isolation of the compound from natural sources or by biotechnological method.

Thousands of flavoring raw materials may be needed by a flavor plant. A large number of the flavoring raw materials are supplied by different manufacturers and stocked. Therefore, a strict quality control system for the flavoring raw materials, as well as the products, is very important.

FLAVOR MANUFACTURING

Flavor Compounding

From thousands of flavor raw materials, twenty to fifty items are commonly selected and mixed with different ratios to blend a favor. This is called flavor compounding, which is a kind of formulation. The raw material may be organic chemicals, essential oils, extracts, oleoresins, or processed flavors. Knowledge of their nature, physical and organoleptic properties, and applications are needed by a flavorist. Doing flavor compounding requires at least 3 to 5 years of training.

How a flavor is formulated and modified is shown by using strawberry flavor as an example (Table 9.1). The characteristic notes of strawberry are fruity, sweet, green, and a little bit oily and sour. Ethyl butyrate and methyl cinnamate have fruity note, cis-3-hexen-1-ol and ethyl hexanoate are green, benzaldehyde and 2,5-methyl-4-hydroxy-3(2H)furanone are sweet, butyric acid is sour, and γ-undecalactone is oily-fruity. Formula 1 was originally designed for use in cake mix. Therefore, compounds of a lower boiling point, such as ethyl acetate, were not used. The solvent used was propylene glycol, which has a boiling point of 187.3°C. So the flavor compound was heat-stable. However, the application of this flavor in cake resulted in a sensory characteristic of pineapple-like, but not strawberry. Pineapple is oily, fruity, and sweet. Since butter, sugar, and milk were among the ingredients in the cake mix, oily or fatty and sweet notes were derived from those ingredients. Therefore, formula 1 has to be modified by reducing the amount of γ-undecalactone, benzaldehyde, and 2,5-dimethyl-4-hydroxy-3(2H)furanone and increasing the amount of ethyl butyrate, ethyl hexanoate, cis-hexen-1-ol, butyric acid, and methyl cinnamate. The

TABLE 9.1. Formulas of Strawberry Flavors.

	Formula 1	Formula 2
Ethyl butyrate	1.5	3.0
Ethyl hexanoate	1.0	2.0
cis-3-Hexen-1-ol	1.0	2.0
Benzaldehyde	0.3	0.2
Butyric acid	0.4	0.8
2,5-Dimethyl-4-hydroxy-3(2H)furanone	2.0	1.0
Methyl cinnamate	1.3	2.0
γ-Undecalactone	0.9	0.5
Propylene glycol	91.6	88.5
	100.0	100.0

application test showed that formula 2 gives the cake a strawberry flavor although this modified flavor did not smell so strawberry before application.

Process Flavor

Process flavors include processed (reaction) flavor, fat flavors, hydrolysates, autolysates, enzyme modified flavor, etc. Production of dairy flavor by enzyme modification of butterfat is a recent example of flavors produced by enzymatic reactions (Lee et al., 1986; Manley, 1994). However, meat flavor produced by enzymatic reactions has a much longer history.

Raw meat has little flavor. Characteristic meat flavor varies with the species of animal and the temperature and type of cooking. Both water-soluble and lipid-soluble fractions of meat contribute to meat flavor. The water-soluble components include precursors which upon heating are converted to volatile compounds described as "meaty." Many desirable meat flavor volatiles are synthesized by heating water-soluble precursors such as amino acids and carbohydrates. The Maillard reaction, including formation of Strecker degradation compounds and interactions between aldehydes, hydrogen sulfide, and ammonia, is important in the formation of the volatile compounds of meat flavor. In addition, other kinds of flavors formed during cooking can also be obtained from heat processing, theoretically. The contingency is the availability of the precursors, which may be too expensive to be isolated from natural raw materials or to be synthesized.

The most practical way to characterize process flavorings is by their starting materials and processing conditions, since the resulting composition of volatiles is extremely complex, comparable to the composition of cooked foods. Process flavorings are produced every day by housewives in kitchens, by the food industries during food processing, and by the flavor industry. The International Organization of the Flavor Industry (IOFI) has guidelines for the production and labeling of process flavorings (IOFI, 1990). Some key points are: the reactants are strictly appointed; flavorings, flavoring substances, flavor enhancers, and process flavor adjuncts shall be added only after processing is completed; the processing conditions should not exceed 15 min at 180°C or proportionately longer at lower temperatures; and the pH should not exceed 8.0.

Process flavors are very successful in some cases, but also unsuccessful in many cases. Natural flavor materials such as meat extract or aromatic chemicals may be added to process flavors to enrich some notes or to raise the flavor intensity.

APPLICATIONS OF FLAVORS

Choosing the right type of flavor, dosage, and adding the flavor in the right step in food processing are important in flavor applications. A flavor can be admired only after suitable application. Due to different application conditions, flavors are made to have different characteristics, i.e., solubility in water or oil, heat-stable or -unstable to meet the requirements. There is no general rule for flavor application. Flavor users should have some basic knowledge of flavor, food chemistry, and processing and then can handle flavor applications work very well. The following are some examples for using the flavor in the right way. Citral is the key compound of lemon flavor. If the flavor has undergone thermal treatment severe enough to let it be oxidized, then the hemiacetal form can be added to replace citral. Limonene is a major constituent in citrus oils, and it has to be removed to prevent off-flavor production in food processing and storage. The extent of evaporation loss of each flavor ingredient is different in food processing. Some food components such as starch, lipids, and proteins can trap flavor compounds and reduce their volatilities. Some foods have their own flavors or off-flavor production. Therefore, modification of flavor formulas is needed to meet the identity of different processed foods. Studies on flavor application for each food product are required to find the right strength, right form, and right step. Technical supports to flavor users are standard services provided by flavor makers.

Flavors can be used to develop new food products. At the 1995 IFT Food EXPO, flavor manufacturers created unique berry flavors that do not exist naturally and are more exciting than ever. May-berry, pepperberry, juneberry, mountainberry, bugleberry, and bellberry were all fabricated by creative flavor manufacturers (Sloan, 1995). Creating new flavors to create new flavor applications is a challenge to flavor industry, thus leading to developing more new and quality food products for quality living.

REFERENCES

Ames, J. M. and Hincelin, O. 1992. "Novel sulfur compounds from heated thiamin and xylose/thiamine model systems," *Progress in Flavor Precursor Studies,* Schreier, P. and Winterhalter, P., eds., Allured Publishing Corporation: Carol Stream, IL, pp. 379–382.

Arctander, S. 1960. *Perfume and Flavor Materials of Natural Origin.* Available from Maria G. Arctander, 6665 Valley View Blvd., Las Vegas, NV 89118.

Arctander, S. 1969. *Perfume and Flavor Chemicals,* Published by the Author. Available from Maria G. Arctander, 6665 Valley View Blvd., Las Vegas, NV 89118.

Block, E. and Calvey, E. E. 1994. "Facts and Artifacts in Allium Chemistry," *Sulfur Compounds in Foods,* Mussinan, C. J. and Keelan, M. E., eds., American Chemical Society: Washington, D.C., p. 63.

Buckholz, L. L. 1988. "Maillard technology as applied to meat and savory flavors," *Thermal Generation of Aromas,* Parliment, Thomas H., McGorrin, Robert J. M., and Ho, Chi-Tang, eds., American Chemical Society, Los Angeles, pp. 406–420.

Buttery, R. G. and Ling, L. 1992. "Enzymatic production of volatiles in tomatoes," *Progress in Flavor Precursor Studies,* Schreier, P. and Winterhalter, P., eds., Allured Publishing Corporation: Carol Stream, IL, pp. 137–146.

Chang, C. H., Chung, C. C. and Wu, C. M. 1991. "Volatile Components of Various Shiitake Products (*Lentinus edodes Sing*)," *Food Science* (Taiwan), 18(3):199–204.

Chen, C. C., Chen, S. D., Chen, J. J. and Wu, C. M. 1984. "Effects of pH value on the formation of volatiles of shiitake (*Lentinus edodes*), an edible mushroom," *J. Agric. Food Chem.,* 32(5):999–1001.

Chen, C. C. and Wu, C. M. 1984. "Studies on the enzymic reduction of 1-octane-3-one in mushroom (*Aragicus bisporus*)," *J. Agric. Food Agric.,* 32(6):1342–1344.

de Kok, P. M. T. and Rosing, E. A. E. 1994. "Reactivity of peptides in the Maillard reaction." *Thermally Generated Flavors: Maillard, Microwave and Extrusion Process,* Parliment, T. H., et al., eds., American Chemical Society, Washington, D.C., pp. 158–179.

de Roos, K. B. 1992. "Meat flavor generation from cysteine and sugars," *Flavor Precursors,* Teranishi, R. et al., eds., American Chemical Society: Washington, D.C., pp. 203–216.

Farmer, L. J. and Whitefield, F. B. 1993. "Aroma compounds formed from the interaction of lipid in the Maillard reaction," *Progress in Flavor Precursor Studies,* Schreier, P. and Winterhalter, P., eds., Allured Publishing Corporation: Carol Stream, IL, pp. 387–390.

Furia, T. E. and Bellanca, N. 1975. *Fenaroli's Handbook of Flavor Ingredients,* 2nd Edition, CRC Press: Ohio, Inc.

Galliard, T. and Matthew, J. A. 1977. "Lipoxygenase-mediated cleavage of fatty acids to carbonyl fragments in tomato fruits," *Phytochemistry,* 16:339–343.

Giese, J. 1994a. "Modern alchemy: use of flavors in food," *Food Technology,* Feb: 106–113, 116.

Giese, J. 1994b. "Spices and seasoning blends: a taste for all seasons," *Food Technology,* April:88–95, 98.

Gunata, Z., Dugelay, I., Sapis, J. C., Baumes, R., and Bayonove, C. 1992. "Role of Enzymes in the Use of the Flavor Potential from Grape Glycosides in Winemaking," *Progress in Flavor Precursor Studies,* Schreier, P. and Winterhalter, P., eds., Allured Publishing Corporation: Carol Stream, IL, pp. 219–234.

Güntert, M., Bertram, H. J., Emberger, R., Hopp, R., Sommer, H. and Werkoff, P. 1992. "New aspects of the thermal generation of flavor compounds from thiamin," *Progress in Flavor Precursor Studies,* Schreier, P. and Winterhalter, P., eds., Allured Publishing Corporation: Carol Stream, IL, pp. 361–378.

Güntert, M., Bertram, H. J., Emberger, R., Hopp, R., Sommer, H. and Werkoff, P. 1994. "Thermal degradation of thiamin (Vitamin B_1)," *Sulfur Compounds in Foods,* Mussinan, C. J. and Keelan, M. E., eds., American Chemical Society: Washington, D.C., pp. 199–223.

Harlander, S. 1994. "Biotechnology for the Production of Flavoring Materials," *Source Book of Flavors,* Reineccius, G., ed., Chapman & Hall: New York & London, pp. 155–175.

Ho, C. T. and Chen, Q. 1994. "Lipids in food flavors, an overview," *Lipids in Food*

Flavors, Ho, C. T. and Hartman, T. G., eds., American Chemical Society: Washington, D.C., pp. 2–14.

Ho, C-T., Oh, Y-C., Zhang, Y. and Shu, C-K. 1992. "Peptides as flavor precursors in mode Maillard reactions," *Flavor Precursors,* Teranishi, R., Takeoka, G. R., and Güntert, M., eds., American Chemical Society: Washington, D.C., pp. 193–202.

Hodge, J. E. 1953. "Chemistry of browning reactions in model systems," *J. Agric. Food Chem.,* 1:928–943.

Hsieh, R. J. and Kinsella, J. E. 1989. "Lipoxygenase generation of specific volatile flavor carbonyl compounds in fish tissues," *J. Agric. Food Chem.,* 37(2):280–286.

Hsu, J. P., Jenf, J. G., and Chen, C. C. 1993. "Identification of a novel S-compound from the interaction of garlic and heated edible oil," *Progress in Flavor Precursor Studies,* Schreier, P. and Winterhalter, P., eds., Allured Publishing Corporation: Carol Stream, IL, pp. 391–394.

Hwang, H. I., Hartman, T. G., Karure, M. V., Izzo, H. V., and Ho. C-T. 1994. "Aroma generation is extruded and heated wheat flour," *Lipids in Food Flavors,* Ho, C. T. and Hartman, T. G., eds., American Chemical Society: Washington, D.C., pp. 144–157.

IOFI. 1990. "IOFI Guidelines for the Production and Labeling of Process Flavorings," *Code of Practice for the Flavor Industry,* 2nd Edition, International Organization of the Flavor Industry: Geneva, pp. F1–F4.

Izzo, H. V., Yu, T-H., and Ho, C-T. 1992. "Flavor generation from the Maillard reaction of Peptides and Proteins," *Progress in Flavor Precursor Studies,* Schreier, P. and Winterhalter, P., eds., Allured Publishing Corporation: Carol Stream, IL, pp. 315–328.

Kim, S. M., Wu, C. M., Kubota, K., and Kobayaski, A. 1995. "Effect of soybean oil on garlic volatile compounds isolated by distillation," *J. Agric. Food Chem.,* 43(2):449–452.

Kobyashi, A., Winterhalter, P., Morita, K., and Kubota, K. 1992. "Glycosides in fresh tea leaves—precursors of black tea flavor," *Progress in Flavor Precursor Studies,* Schreier, P. and Winterhalter, P., eds., Allured Publishing Corporation: Carol Stream, IL, pp. 257–260.

Kuo, J. M. and Pan, B. S. 1991. Effect of lipoxygenase on formation of cooked shrimp flavor compound—5, 8, 11-tetradecatrine-2-one. *Agric. Biol. Chem.,* 55:847–848.

Kuo, M. C., Chen, S. L., Wu, C. M., and Chen, C. C. 1985. "Changes in volatile components of passion fruit juice as affected by centrifugation and pasteurization," *J. Food Sci.,* 50(4):1208–1210.

Kuo, J. M., Pan, B. S., Zhang, H. and German, J. B. 1994. "Identification of 12-lipoxygenase in the hemolymph of tiger shrimp (*Penaeus japonicus Bate*)," *J. Agri. Food Chem.,* 42:1620–1623.

Lawrence, B. M. 1986. "Essential Oil Production: A Discussion of Influencing Factors," *Biogeneration of Aromas.* Parliment, T. H. and Croteau, R., eds., American Chemical Society: Washington, D.C., pp. 363–369.

Lee, K. M., Shi, H., Huang, A.-S., Carlin, J. T., Ho, C.-T., Chang, S. S. 1986. "Production of a romano cheese flavor by enzymic modification of butterfat," *Biogeneration of Aromas,* Parliment, T. H. and Croteau, R., eds., American Chemical Society: Washington, D.C. pp. 370–378.

Manley, C. 1994. "Process flavors," *Source Book of Flavors,* Reineccius, G., ed., Chapman & Hall: New York, London, pp. 139–154.

Min, D. B. and Smouse, T. H. 1989. *Flavor Chemistry of Lipid Foods,* American Chemists' Society, Champaign, IL.

Moller, A. B. 1981. "Chemical changes in ultra heat treated milk during storage," *Maillard Reactions in Food,* Ericksson, C., ed., Pergamon Press: Oxford, UK. p. 357.

Moshonas, M. G. and Shaw, P. E. 1987. "Flavor evaluation of fresh and aseptically packaged orange juice," *Frontiers of Flavor,* Charalambous, G., ed., Elsevier Science Publishers B. V.: Amsterdam, pp. 133–145.

Mottram, D. S. and Salter, L. J. 1988. "Flavor formation in meat-related Maillard systems containing phospholipids," *Thermal Generation of Aromas,* Parliment, Thomas H., McGorrin, Robert J. M., and Ho, Chi-Tang, eds., American Chemical Society, Los Angeles, pp. 442–451.

Nagy, S. and Shaw, P. E. 1990. "Factors Affecting the Flavor of Citrus Fruit," *Food Flavours: Part C. The Flavor of Fruits.* Morton, I. D. and Macleod, A. J., eds., Elsevier Science Publishers: Amsterdam-Oxford-New York-Tokyo, pp. 93–124.

Ouweland, G. A. M., Demole, E. P., and Enggist, P. 1988. "Process meat flavor development and the Maillard reaction," *Thermal Generation of Aromas,* Parliment, Thomas H., McGorrin, Robert J. M., and Ho, Chi-Tang, eds., American Chemical Society, Los Angeles, pp. 433–441.

Pan, B. S. and Kuo, J. M. 1994. "Flavor of Shellfish and Kamboko Flavorants," *Seafoods: Chemistry, Processing Technology and Quality.* Shahidi, F. and Botta, J. R., eds., Blackie Academic and Professional: Glasgow, pp. 85–114.

Pangborn, R. M. 1981. "A Critical Review of Threshold, Intensity and Descriptive Analysis in Flavor Research," *Flavor '81.* Schreier, P., ed., Walter de Gruyter: Berlin and New York, pp. 1–32.

Pearson, A. M. and Gray, J. I. 1983. "Mechanism responsible for warmed-over flavor in cooked meat," *The Maillard Reaction in Foods and Nutrition,* Waller, G. R. and Feather, M. S., eds., American Chemical Society: Washington, D.C., pp. 287–300.

Pruthi, J. S. 1980. *Spices and Condiments: Chemistry, Microbiology, Technology.* London: Academic Press, p. 44.

Reineccius, G. 1994a. "The Flavorist," *Source Book of Flavors,* Reineccius, G., ed., Chapman & Hall: New York, London, pp. 691–712.

Reineccius, G. 1994b. "Natural flavoring materials." *Source Book of Flavors.* Reineccius, G., ed., Chapman & Hall: New York, London, pp. 176–364.

Reineccius, G. 1994c. *Source Book of Flavors.* Reineccius, G., ed., New York, London: Chapman & Hall, p. 63.

Sadafian, A. and Crouzet, J. 1987. "Aroma compounds retention during extrusion cooking," *Frontiers of Flavor,* Charalambous, G., ed., Elsevier Science Publishers B. V.: Amsterdam, pp. 623–637.

Salinas, J. P., Hartman, T. G., Karmas, K., Lech, J. and Rosem, R. T. 1994. "Lipid-derived aroma compounds in cooked potatoes and reconstituted dehydrated potato granules," *Lipids in Food Flavors,* Ho, C. T. and Hartman, T. G., eds., American Chemical Society: Washington, D.C., pp. 108–129.

Saltmarch, M., Vagnini-Ferrari, M. and Labuza, T. P. 1981. "Theoretical basis and application of kinetics to browning in spray-dried whey food systems," *Maillard Reactions in Food,* Eriksson, C., ed., Pergamon Press: Oxford, UK, pp. 331–344.

Schreier, P. and Lorenz, G. 1981. "Formation of 'Green-Grassy'—Notes in Disrupted

Plant Tissues: Characterization of the Tomato Enzyme Systems," *Flavour '81,* Schreier, P., ed., Berlin and New York: Walter de Gruyter, pp. 495–507.

Schreier, P. and Winterhalter, P. 1986. "Precursors of papaya (*Carica papaya,* L.) fruit volatiles," *Biogeneration of Aromas,* Parliment, T. H. and Croteau, R., eds., American Chemical Society: Washington, D.C., pp. 85–98.

Shi, H. and Ho, C. T. 1994. "The flavor of poultry meat," *Flavor of Meat and Meat Products,* Shahidi, F., ed. Blackie Academic and Processional: London. pp. 52–70.

Shibamoto, T. and Yeo, H. 1992. "Flavor compounds formed from lipids by heat treatment," *Flavor Precursors Thermal and Enzymatic Conversions,* Teranishi, R. et al. eds., American Chemical Society: Washington, D.C., pp. 175–182.

Sloan, A. E. 1995. "Ingredients add more fun, flavor, freshness & nutrition," *Food Technology,* Aug. pp. 102, 16.

Straus, C. R., Wilson, B., Gooley, P. R. and Williams, P. J. 1986. "Role of Monoterpenes in Grape and Wine Flavor," *Biogeneration of Aromas,* Parliment, T. H. and Croteau, R., eds., American Chemical Society: Washington, D.C. pp. 222–242.

Takeoka, G. R., Flath, R. A., Buttery, R. G., Winterhalter, P., Guntert, M., Ramming, D. W. and Teranishi, R. 1992. "Free and bound flavor constituents of white-fleshed nectarines," *Flavor Precursors Thermal and Enzymatic Conversions,* Teranishi, R., Takeoka, G. R., and Guntert, M., eds., American Chemical Society: Washington, D.C., pp. 116–138.

Teranishi, R., Buttery, R. G., and Guadagni, D. G. 1981. "Some Properties of Odoriferous Molecules," *Flavor '81.* Schreier, P., ed., Berlin and New York: Walter de Gruyter, pp. 133–143.

Tressl, R., Daoud, B., and Engel, K. H. 1982. "Formation of eight-carbon and ten-carbon components in mushrooms (*Agaricus campestris*)," *J. Agric. Food Chem.,* 30(1):89–93.

Villota, R. and Hawkes, J. G. 1988. "Flavoring in extrusion, an overview," *Thermally Generated Flavors: Maillard, Microwave, and Extrusion Processes,* Parliment, T. H., ed., American Chemical Society: Washington, D.C., pp. 280–295.

Wu, C. M. and Chen, S. Y. 1992. "Volatile compounds in oils after deep frying or stir frying and subsequent storage," *JAOCS,* 69(9):858–865.

Wu, C. M. and Liou, S. E. 1986. "Effect of tissue disruption on volatile constituents of bell peppers," *J. Agric. Food Chem.,* 34(4):770–772.

Wu, C. M. and Liou, S. E. 1992. "Volatile components of water-boiled duck meat and cantonese style roasted duck," *J. Agric. Food Chem.,* 40(5):838–841.

Wu, C. M., Liou, S. E. and Chiang, W. 1987. "Volatile compounds of the wax gourd (*Benincasa hispida,* Cogn) and a wax gourd beverage," *J. Food Sci.* 52(1):132–134.

Wu, C. M., Liou, S. E., and Wang, M. C. 1986. "Changes in volatile constituents of bell peppers immediately and 30 minutes after stir frying," *JAOCS,* 63(9):1172–1175.

Wurzenberger, M. and Grosch, W. 1982. "The enzymic oxidative breakdown of linoleic acid in mushrooms (*Psalliota bispora*)," *Z. Lebensm. Unters Forsch.,* 175:186–190.

Yamaguchi, M. and Wu, C. M. 1975. "Composition and Nutritive Value of Vegetables for Processing," *Commercial Vegetable Processing,* Luh, B. S. and Woodroof, J. G., eds., Westport, CT: The AVI Publishing Company, Inc., pp. 652–653.

Yu, T. H. and Wu, C. M. 1989. "Stability of allicin in garlic juice," *J. Food Sci.* 54(4):977–981.

Yu, T. H., Wu, C. M., and Ho, C. T. 1993. "Volatile compounds of deep-oil fried, microwave-heated, and oven-baked garlic slices," *J. Agric. Food Chem.,* 41(5):800–805.

Yu, T. H., Wu, C. M., and Ho, C. T. 1994. "Meat-like flavor generated from thermal interactions of glucose and alliin or deoxyalliin," *J. Agric. Food Chem.* 42(4):1005–1009.

Yu, T. H., Wu, C. M., and Liou, Y. C. 1989. "Volatile compounds from garlic," *J. Agric. Food Chem.,* 37(3):725–730.

CHAPTER 10

Main Functional Food Additives

ADRIAAN RUITER
ALPHONS G. J. VORAGEN

INTRODUCTION

THE addition of certain substances not intended as functional or sensory ingredients to foodstuffs was practiced in ancient times, mostly for improving keeping properties. Salt was added to perishable foodstuffs such as meat and fish, from prehistoric ages onwards. Smoke-curing can also be considered as the fortuitous addition of constituents to food, as wood smoke contains a number of compounds that are absorbed by the food during the smoke-curing process or are deposited onto the surface.

The preparation of any food product includes the addition of a number of ingredients that are not considered to be additives, but that clearly improve some properties of the food, such as keeping quality, and are originally intended as such. Preparation of a marinade in sour wine or vinegar, for example, is a technique for preserving fish, which was already known to the Romans, but acetic acid is not an additive in the strict sense of the word. In some cases, it is not so easy to determine whether or not the substance under consideration is an additive. It is helpful, however, to keep in mind that an additive is intended as an *aid,* for some purpose or another, and not as an *ingredient.*

In 1955, the joint FAO/WHO Expert Committee on Nutrition defined food additives as "non-nutritive substances which are intentionally added to foodstuffs, mostly in small quantities, with the aim of improving the appearance, the flavour, the taste, the composition or the shelf-life." In a more recent wording, these food additives are described as "substances generally unintended as a foodstuff or as a characteristic ingredient of a foodstuff which, irrespective of any nutritional value, are added, for any technological or sensoric reason, to a foodstuff during manufacturing, preparation, packaging, transport or storage, and from which it is expected that either the substance itself or reaction or decomposition products become a permanent component of the foodstuff or the raw material" (van Dokkum, 1985; Kamsteeg and Baas, 1985). In the latter definition, the term *improvement* is not included, and the remaining presence of the compound, or reaction products from that compound, are included in the definition.

This may be the result of a shift in the attitude towards additives. Concern about these components is sometimes stimulated by press coverage in which their safety is questioned. The perceived ranking, by the public, of hazards prevailing in food consumption (Hall, 1971) is considerably different from the ranking of actual hazards (Wodicka, 1977). Contrary to this, food additives have been subject to extremely careful laboratory screening before they are used (Pilnik and Folstar, 1979), and the risks induced by food additives are regarded as very small (Doll and Peto, 1981).

The origin of food additives often remains a point of discussion. There is a continuing demand, from the consumer's side, for "natural" additives. No additive, however, is completely free from impurities. Products of chemical synthesis should be purified, eliminating starting materials and compounds resulting from side reactions. "Natural" compounds should be purified as well in order to remove accompanying substances that have no significance in the final product. Generally speaking, purification is more difficult and more complicated for "natural" additives, as it is also much more problematic to characterize the raw material, which may contain a great many of ill-defined compounds whose toxicity is largely unknown. In products of chemical synthesis, the presence of ill-defined and toxic components cannot be completely excluded either, but, as a rule, the situation is not characterized by such an extreme complexity (Ruiter, 1989). Feberwee (1989) points out that official legislation does not discriminate between safe natural and safe artificial food additives. The main difference in safety evaluation, between these two categories, is the long experience of man with natural additives (Lüthy, 1989).

CLASSIFICATION

Additives are mostly listed and classified into categories, i.e.:

- preservatives to extend the shelf-life of foodstuffs
- antioxidants to protect lipids in food from attack by oxygen
- colorants to improve the appearance of a foodstuff
- emulsifying agents to enable a fine partition of oils or fats in water (or a partition of water in oils or fats)
- stabilizers to prevent breaking of emulsions

- thickening and gel-forming agents
- clarifying agents
- substances improving the nutritional value
- substances improving organoleptic properties
- flavor enhancers to improve the perception of taste and flavors
- glazing agents
- sweeteners to replace sugars in giving a sweet taste to the product
- many other substances such as anti-clotting agents, moisteners, anti-foaming agents, flour improvers, leavening agents and baking powders, melting salts, stiffening agents, complexing agents, fillers, enzymes, and so on.

With respect to reactivity of additives, it is preferable to make another classification, in which three groups can be distinguished:

(1) Substances that, simply by their presence, lead to the desired improvement. Most colorants, sweeteners, vitamins, and some preservatives belong to this class. Many of these additives do not display a strong reactivity towards other constituents during preparation and storage of the foodstuffs to which they are added.

(2) Substances that are added because of their reactivity towards undesirable components already present or arising during manufacturing or storage and that may be bound by these added substances. The reaction may be directed towards these components themselves or towards their precursors. Antioxidants are an example of this category, as well as some peculiar substances reacting with matrix components to make desirable components such as flavor compounds.

(3) Food additives that participate, for a part, in fortuitous reactions, which, in some cases, may be undesirable

This classification, however, also has its limits. First, there is hardly any additive that does not take part in some chemical reaction at all. Furthermore, some food additives may also participate in unintentional reactions. Therefore, in this presentation, some additives are discussed in an individual way, with emphasis on their reactivity towards matrix compounds.

PRESERVATIVES

INTRODUCTION

Preservatives are added to a variety of foodstuffs in order to protect them against microbial spoilage. This protection is possible, in many cases, because of a chemical reaction between the preservative and the microorganisms. It may therefore be expected that these compounds show some reactivity towards food components as well.

Many preservatives are comparatively reactive compounds, e.g., bisulfite, nitrate, sorbic acid, but some of these show a moderate reactivity only, e.g., benzoic acid. Bisulfite, nitrite, and sorbic acid are discussed below.

SULFITE

Disinfection by the vapors of burning sulphur is an old technique that was frequently used to decontaminate wine casks. Some sulphur dioxide was left in these vessels, preventing the wine from unwanted microbial infections.

At present, sulfites are used both as preservatives and as agents that stop browning reactions. In food, the HSO_3^- species predominates while, in dehydrated food, it is expected that S(IV) mainly exists as metabisulfite ($S_2O_5^-$), which is in equilibrium with HSO_3^- and SO_3^- (Wedzicha et al., 1991). Because of the nucleophilicity of the sulfite ion, many reactions with food components are possible (Wedzicha, 1991), one of these being the reversible addition to carbonyl compounds. It is suggested that the sulfite, rather than the bisulfite, ion acts as the nucleophilic agent (Wedzicha et al., 1991).

This reaction has many implications for foodstuffs. Aroma components possessing a carbonyl group, for example, become involatile and do not contribute anymore to the overall flavor. Other nucleophilic reactions include the cleavage of $S-S$ bonds in proteins and addition to $C=C$ bonds of α,β-unsaturated carbonyl compounds. Control of nonenzymatic browning is based upon this latter reastion (McWeeny et al., 1974). A key intermediate of the Maillard reaction, i.e., 3,4-deoxyhexulos-3-ene, is efficiently blocked by a fast reaction with sulfite, leading to formation of 3,4-dideoxy-4-sulphohexosulose, which is much less reactive and in which sulfite is irreversibly bound.

Ascorbic acid browning is also inhibited by the addition of sulfite (Wedzicha and McWeeny, 1974). The same holds for polyphenol oxidase-catalyzed oxidation of natural phenols in fruit. The mechanism of the inhibition is by reaction of o-quinone intermediates with sulfite, which leads to nonreactive sulphocatechols (Wedzicha, 1995).

An undesirable reaction of sulfite in food is the cleavage of thiamin by means of an attack at the pyrimidin moiety (Zoltewicz et al., 1984). This was one of the reasons for a ban, in many countries, on the use of sulfite in meat. Another reason is the preserving effect on the meat color, which makes stale meat look as if it were fresh. Sulfite, however, is unable to reduce metmyoglobin back to myoglobin (Wedzicha and Mountfort, 1991).

An important reaction, in a quantitative respect, is the cleavage of disulfide bonds in meat proteins, in particular in lean meat.

The reducing capacities of sulfite should be emphasized as well. In fact, cleavage of $S-S$ bonds by sulfite can be considered as a reduction. This property of sulfite makes it useful as an additive to flour for biscuit making (Stevens, 1966, 1973; cited by Wedzicha, 1995). The cleavage of disulfide bonds in wheat proteins speeds up and facilitates the production of a satisfactory dough.

A quite different type of reaction, which also may occur in food, is that of reduction of azo dyes to colorless hydrazo compounds. Like in the reaction with carbon compounds, the reactive species is SO_3^- and not HSO_3^- (Wedzicha and Rumbelow, 1981).

NITRITE

It has been known for a long time that small amounts of saltpeter (KNO_3) are able to cause a reddish discoloration of meat, which is characteristic for many meat products. About 1890, it was found that it was not nitrate, but its reduction product, i.e., nitrite, that was responsible for this color development.

Some decades later, nitrite was recognized as a potent inhibitor of microorganisms, including pathogens, in many meat products. In particular, the inhibition of *Clostridium botulinum,* with accompanying toxin formation, was established. The role of nitrite in the characteristic cured meat flavor was still noticed later on. This has been discussed in Chapter 8 of this volume.

Fat also is able to bind some nitrite. The amount of nitrite incorporated in fat is considerably higher in unsaturated, than in saturated, lipids.

Ascorbic acid promotes depletion of nitrite. Curiously enough, some nitrite is oxidized to nitrate, and some nitrite is turned into nitrous and nitric oxide and elementary nitrogen. Nitrite is able to react with intermediates of the Maillard reaction such as 3-deoxyosulose (Wedzicha and Wei Tian, 1989) and even with residues of veterinary drugs in meat (Smit et al., 1990).

The reaction of nitrite with secondary or tertiary amines, though unimportant in quantitative respect, leads to N-nitroso compounds, which, for a considerable part, are potent carcinogens. These compounds may rearrange to form highly electrophilic diazonium ions that react with cellular nucleophiles such as water, proteins, and nucleic acids. The carcinogenicity of N-nitroso compounds is thought to result primarily from a reaction of the diazonium ions with various nucleophilic sites of DNA bases (Tiedink, 1991).

Nitrate ingested with food or drinking water is partially reduced to nitrite in the body and contributes more than nitrite in meat products to the possible endogenous formation of N-nitroso compounds. A ban on the use of low amounts of nitrite as an additive is therefore not very rational and deprives the consumer of a very effective guard against a number of pathogenic microorganisms, in particular *Cl. botulinum,* but also *Cl. perfringens* and *Staphylococcus aureus.*

Finally, nitrite may react in physiological concentrations and under gastric pH conditions with naturally occurring, as well as synthetic, antioxidants (Kalus et al., 1990). There are no indications for the formation of hazardous reaction conditions from a viewpoint of mutagenicity.

SORBIC ACID

The preserving properties of sorbic acid were recognized around 1940. During the late 1940s and the 1950s, sorbic acid became available on a commercial scale, resulting in its extensive use as a food preservative throughout the world (Sofos and Busta, 1983). As a straight-chain *trans-trans* diunsaturated fatty acid ($CH_3-CH=CH-CH=CH-COOH$), it is susceptible to nucleophilic attack (Khandelwal and Wedzicha, 1990b). The lowest electron density is associated with position 3. The much greater extent to which the charge on the intermediate arising from attack at position 5 (adjacent to the terminal methyl group) is delocalized, however, suggests that this should be preferred, despite the lower electron density at position 3. Nucleophilic groups, e.g., the thiol group, may bind to the carbon atom at the 5 position (Khandelwal and Wedzicha, 1990a; Wedzicha, 1995).

Sorbic acid is easily oxidized. This oxidation is accompanied by the development of a glyoxal-like flavor in sorbic acid preparations and a brown color in a wide variety of model foods in which sorbic acid was included. Amino acids accelerate color development (Wedzicha et al., 1991).

ANTIOXIDANTS

Antioxidants can be defined as "substances that, when present in low concentrations compared to those of an oxidizable substrate, significantly delay or inhibit oxidation of that substrate" (Halliwell and Gutteridge, 1989).

Antioxidants are frequently added to unsaturated fats and oils in order to protect these against oxidative deterioration. For this reason, they are also added to a variety of food products containing unsaturated lipids. Antioxidants frequently applied are esters of gallic acid, butylated hydroxyanisole (BHA), butylated hydroxytoluene (BHT), and tertiary butyl hydroxyquinone (TBHQ). Of these, TBHQ is by far the most potent antioxidant. In both BHA and BHT, the butyl groups are also of a tertiary structure.

Antioxidants are naturally present in many foodstuffs and are of great importance as inactivators of radical formation. Some antioxidative enzyme systems are produced in the human body and are supposed to play an important role in the cellular defense against oxidative damage (Langseth, 1995).

Many of the antioxidants present in food have the function to terminate chain reactions. A variety of compounds such as phenols, aromatic amines, and conjugates can function as chain-breaking antioxidants. They react with the chain-propagating radical species, which results in the formation of radical species incapable of extracting hydrogen atoms from unsaturated lipids. These radicals may rapidly combine with other radicals or, if a polyphenolic structure is present such as in gallic acid esters, disproportionate into their original state and a quinoid form.

Since there are synergistic effects between antioxidants, commercial preparations usually contain mixtures of these antioxidants. As oxidative rancidity is strongly catalyzed by some heavy metal ions, in particular Cu^{2+}, antioxidant mixtures often contain sequestrants (e.g., citric acid, EDTA) in order to complex these ions. Reductants such as ascorbic acid, which decrease the local concentration of oxygen, are also able to decrease the formation of peroxy radicals.

Fat oxidation by bacteria can be suppressed by the addition of preservatives such as benzoic acid or sorbic acid.

STABILIZERS, EMULSIFIERS, AND THICKENING AGENTS

The most important representatives of these compounds are polysaccharides, e.g., starch and starch derivatives (α-1,4 D-glucans), cellulose and cellulose derivatives (β-1,4 D-glucans), plant extracts (pectins: α-1,4 D-galacturonans), seaweed extracts (carrageenan, agar, alginates), seed flour (guar and locust bean galactomannans, tamarind xyloglucans, konjac glucomannans), exudate gums (arabic, karaya, tragacanth), and microbial gums (xanthan, gellan). Polymers are built up of one or more types of sugar residues, covalently attached in linear, linearly branched, and branched structures. Their anomeric form (α/β), types of linkages (1,2; 1,3; 1,4; 1,4,6; 1,3,6; etc.), presence of functional groups (carboxyl, phosphate, sulphate, esters, ethers), and molecular weight distribution determine their conformation in aqueous systems (stiff/rod-like, random coil, helices), the intra and intermolecular interactions between molecules (dimerization, association, ionic interactions, hydrophobic interactions), and the interactions with other molecules (other polysaccharides, proteins, lipids). These interactions are the basis for viscous behavior, gelling, water binding, film forming, bulking, stabilizing, and emulsifying properties. Some polysaccharides act synergistically in imparting these functions, e.g., locust bean gum and carrageenan, locust bean gum and xanthan, pectin and alginate. Important parameters for applications are pH, heat and shear stability, syneresis properties, shelf life, and compatibility with other food constituents. Derivatives of these polysaccharides with improved functional properties are also used.

Emulsifiers are amphophilic compounds that concentrate at oil/water interfaces, causing a significant lowering of interfacial tension and a reduction in the energy needed to form emulsions. They can be anionic, cationic, and nonionic compounds that have one or more of the following characteristics: surface active, viscosity enhancer, solid absorbent, or hydrophilic/lipophilic balance (HLB). They are added to food emulsions to increase emulsion stability and to attain an acceptable shelf life. Polysaccharides are not surface-active agents but, rather, macromolecular stabilizers that generally function through enhancement of viscosity and enveloping oil droplets in oil in water emulsions. The emulsifiers being used in food manufacture were categorized by Artz (1990) (cf. Table 10.1). Only lecithin is of natural origin. Its main source is soybean, but it is also present in corn, sunflower, cottonseed, rapeseed, and eggs.

Emulsifiers stabilize emulsions in various ways. They reduce interfacial tension and may form an interfacial film that prevents coalescence of droplets. In addition, ionic emulsifiers provide charged groups on the surface of the emulsion droplets and, thus, increase repulsive forces between droplets. Emulsifiers also can form liquid crystalline microstructures such as micelles at the interface of emulsion droplets. These are formed only at emulsifier concentrations larger than the critical micelle-forming concentration. These microstructures have a stabilizing effect.

The selection of emulsifiers to prepare food emulsions is mainly based on their HLB number. This index expresses the hydrophile-lipophile balance and is based on the relative percentage of hydrophilic to lipophilic groups within the emulsifier molecule. Lower HLB numbers indicate a more lipophilic emulsifier, while higher numbers indicate a more

TABLE 10.1. **Food Emulsifier Categories.**

Category	Typical Application
Lecithin (naturally occurring) and lecithin derivatives	
Glycerol fatty acid esters	
Hydroxycarboxylic acid and fatty acid esters	Baking goods, margarine
Lactylate fatty acid esters	
Polyglycerol fatty acid esters	Baked goods
Polyethylene and propylene glycol fatty acid esters	O/W emulsions
Ethoxylated derivatives of monoglycerides	
Sorbitan fatty acid esters	Antistaling

hydrophilic emulsifier. Emulsifiers with HLB numbers between 3 and 6 are best for water in oil emulsions and emulsifiers with numbers between 8 and 18 are best for oil in water emulsions.

CLARIFYING AGENTS AND FILM FORMERS

Clarifying agents or flocculants are used to eliminate turbidity or suspended particles from liquids, e.g., chill haze in beer, precipitates in fruit juices and wines, or haze in oils. Often, they provide a nucleation site for suspended fines. Examples of clarifying agents are lime in sugar juice clarification, pectic enzymes to break down pectins in fruit juices, and gelatin for clarification of fruit juices.

Film formers are used to coat a food by providing it with a protective layer and so make it more attractive in appearance or to increase its palatability. Film formers may not impart flavor or mouthfeel of their own to the food. Examples are starches to coat proteins to prevent Maillard reactions, mineral oils to seal pores of eggs, or sodium caseinate to encapsulate fat in whiteners.

FLAVORINGS, COLORANTS, AND SWEETENERS

Artificial flavorings are frequently added to a variety of foodstuffs. These preparations mostly consist of a large number of different compounds of which some show a considerable reactivity. The way in which flavor components interact with the food matrix and how this influences flavor perception has been recently reviewed by Bakker (1995).

Many interactions are of a pure chemical nature and may result from the presence of aldehydes and their reactivity towards amino and thiol groups of proteins. Another frequently occurring type of interaction is the formation of hydrogen bonds between food compounds and polar flavor components such as alcohols. Starch is able to form inclusion complexes with many flavor components. Many other interactions, although of great influence on flavor perception, are of a physical nature and therefore are not mentioned in this chapter.

Food additives such as dyes and sweeteners are not intended to react with matrix compounds or to undergo other reactions. Some reactions may occur, however, and two examples are given here.

As for food dyes, many of these are azo compounds, which implies the possibility of reduction, e.g., by the action of certain bacteria. The loss of color, in these cases, is an indication of spoilage. Bisulfite is also able to reduce azo dyes (Wedzicha and Rumbelow, 1981).

Another example is the sweetener aspartame, which is the methyl ester of N-L-α-aspartyl-L-phenylalanine. Because of its nature, the stability in aqueous systems is limited. The maximum stability is between pH 3 and 5 and decreases at higher temperatures with concomitant loss of sweetening power. The main degradation product is 3,6-dioxo-5-(phenylmethyl)-2-piperazinoacetic acid (Furda et al., 1975). Other decomposition products were listed by Stamp and Labuza (1989), who added some novel components to this group. These all have in common that a sweet taste is absent. Apart from this, aspartame shows remarkable reactivity towards a number of aldehydes that may be present in foodstuffs and contribute to flavor (Hussein et al., 1984; Cha and Ho, 1988).

Many unintended and sometimes unwanted reactions of artificial dyes, sweeteners, and other additives with the food matrix are imaginable and should always be taken into consideration when the consequences of such additions to food are discussed.

REFERENCES

Artz, W. (1990) Emulsifiers. In: *Food Additives* (A.L. Bramen, P. M. Davidson and S. Salminen, eds.), Marcel Dekker Inc., New York, pp. 347–393.

Bakker, J. (1995) Flavor interactions with the food matrix and their effects on perception. In: *Ingredient interactions — Effects on food quality,* A. G. Goanker, ed., Marcel Dekker Inc., New York/Basel, pp. 411–439.

Cassens, R. G., Woolford, G., Lee, S. H. and Goutefongea, R. (1976) Fate of nitrite in meat. In: *Proc. 2nd Int. Symp. Nitrite in Meat Products,* Zeist, the Netherlands. B. Krol and B. J. Tinbergen, eds. Pudoc, Wageningen, pp. 95–100.

Cha, A. S. and Ho, C. T. (1988) Studies of the interaction between aspartame and flavor vanillin by high-performance liquid chromatography. *J. Food Sci.,* 53:562–564.

Doll, R. and Peto, R. (1981) The causes of cancer: quantitative estimates of avoidable risks of cancer in the United States today. *J. Nat. Cancer Institute,* 66:1191–1308.

Feberwee, A. (1989) Legal aspects of food additives of natural origin. In: *Proceedings of the International Symposium Food Additives of Natural Origin,* Plovdiv, Bulgaria, 31 May–2 June 1989, pp. 22–34.

Furda, I., Malizia, P. D., Kolor, M. G. and Vernieri, P. J. (1975) Decomposition products of L-aspartyl-L-phenylalanine methyl ester in various food products and formulations. *J. Agr. Food Chem.* 23:340–343.

Hall, R. L. (1971) Information, confidence, and sanity in the food sciences. *The Flavour Industry,* 2:455–459.

Halliwell, B. and Gutteridge, J. M. C. (1989) *Free radicals in biology and medicine,* 2nd Ed., Clarendon Press, Oxford.

Hussein, M. M., D'Amelia, R. P., Manz, A. L., Jacin, H. and Chen, W.-T. C. (1984) Determination of reactivity of aspartame with flavor aldehydes by gas chromatography, HPLC and GPC. *J. Food Sci.,* 49:520–524.

Joint FAO/WHO Expert Committee on Nutrition (1955) 4th report. *WHO Technical Report Series, No. 97.* World Health Organization, Geneva, pp. 29–33.

Kalus, W. H., Münzner, R. and Filby, W. G. (1990) Isolation and characterization of some products of the BHA-nitrite reaction: examination of their mutagenicity. *Food Additives and Contaminants,* 7:223–233.

Kamsteeg, J. and Baas, M. I. A. (1985) *E = eetbaar.* H. J. W. Becht, Amsterdam.

Khandelwal, G. D. and Wedzicha, B. L. (1990a) Derivatives of sorbic acid-thiol adducts. *Food Chemistry,* 37:159–169.

Khandelwal, G. D. and Wedzicha, B. L. (1990b) Nucleophilic reactions of sorbic acid. *Food Additives and Contaminants,* 7:685–694.

Ladikos, D. and Lougovois, V. (1990) Lipid oxidation in muscle foods: a review. *Food Chemistry,* 35:295–314.

Langseth, L. (1995) *Oxidants, antioxidants, and disease prevention.* ILSI Europe Concise monograph series. ILSI Europe, Brussels, 24 pp.

Lüthy, J. (1989) Safety evaluation of natural food additives. In: *Proceedings of the International Symposium Food Additives of Natural Origin,* Plovdiv, Bulgaria, 31 May–2 June 1989, pp. 35–40.

McWeeny, D. J., Knowles, M. E. and Hearne, J. F. (1974) The chemistry of non-enzymic browning in foods and its control by sulphur. *J. Sci. Food Agric.,* 25:735–746.

Möhler, K. (1973) Formation of curing pigments by chemical, biochemical or enzymatic reactions. In: *Proc. Int. Symp. Nitrite in Meat Products,* Zeist, the Netherlands. B. Krol and B. J. Tinbergen, Eds., Pudoc, Wageningen, pp. 13–19.

Pilnik, W. and Folstar, P. (1979) Entwicklungstendenzen in der Lebensmitteltechnologie. *Deutsche Lebensmittel-Rundschau,* 75:235–248.

Ruiter, A. (1989) Safety of food: the vision of the chemical food hygienist. In: J. P. Roozen, F. M. Rombouts and A. G. J. Voragen, Eds., *Food science: basic research for technological progress.* Proceedings of the symposium in honour of Professor W. Pilnik, Wageningen, the Netherlands, 25 November, 1988, pp. 19–28.

Smit, L. A., Haagsma, N., Hoogenboom, L. A. P. and Berghmans, M. C. J. (1990) Transformation of sulphadimidine in raw fermented sausages. In: *Proc. EuroResidue,* Noordwijkerhout, the Netherlands. N. Haagsma, A. Ruiter and P. B. Czedik-Eysenberg, eds., pp. 346–350.

Sofos, J. N. and Busta, F. F. (1983) In: *Antimicrobials in foods.* A. L. Branen and P. M. Davidson, eds., Marcel Dekker Inc., New York/Basel, p. 141.

Stamp, J. A. and Labuza, T. P. (1989) Mass spectrometric determination of aspartame decomposition products: Evidence for β-isomer formation in solution. *Food Additives and Contaminants,* 6:397–414.

Tiedink, H. G. M. (1991) Occurrence of indole compounds in some vegetables; toxicological implications of nitrosation with emphasis on mutagenicity. Thesis, Wageningen, p. 15.

van Dokkum, W. (1985) Additieven en contaminanten. *Voeding in de praktijk,* 6:1.

Wedzicha, B. L. (1991) Sulphur dioxide—the most versatile food additive? *Chemistry in Britain,* 1030–1032.

Wedzicha, B. L. (1995) Interactions involving sulfites, sorbic acid, and benzoic acid. In: *Ingredient interactions—Effects on food quality,* A. G. Goankar, ed., Marcel Dekker Inc., New York/Basel, pp. 529–559.

Wedzicha, B. L., Bellion, I. and Goddard, S. J. (1991) Inhibition of browning by sulfites. In: *Nutritional and toxicological consequences of food processing*. M. Friesman ed., Plenum Press, pp. 217–236.

Wedzicha, B. L. and McWeeny, D. J. (1974) Non-enzymic browning of ascorbic acid and their inhibition. The production of 3-deoxy-4-sulphopentosulose in mixtures of ascorbic acid, glycine and bisulphite ion. *J. Sci. Food Agric.*, 25:577–587.

Wedzicha, B. L. and Mountfort, K. A. (1991) Reactivity of sulphur dioxide in comminuted meat. *Food Chemistry*, 39:281–297.

Wedzicha, B. L. and Rumbelow, S. J. (1981) The reaction of an azo food dye with hydrogen sulphite ions. *J. Sci. Food Agric.*, 32:699–704.

Wedzicha, B. L., Rimmer, Y. L. and Khandelwal, G. D. (1991) Catalysis of Maillard browning by sorbic acid. *Lebensmittel-Wiss. u. -Technol.*, 24:278–280.

Wedzicha, B. L. and Wei Tian (1989) Kinetics of the reaction between 3-deoxyhexulose and nitrite ion. *Food Chemistry*, 31:189–203.

Wodicka, V. O. (1977) Food safety—rationalizing the ground rules for safety evaluation. *Food Technology*, 31(9):75–77, 79.

Zoltewicz, J. A., Kauffman, G. M. and Uray, G. (1984) A mechanism for sulphite ion reacting with vitamin B_2 and its analogues, *Food Chemistry*, 15:75–91.

CHAPTER 11

Food Safety

JULIE MILLER JONES

INTRODUCTION

SAFE food—it's what every individual expects in every mouthful and every government desires for its populous. Since food is the object of our earliest preferences and the subject of our strongest prejudices, food safety is a gut issue. Yet what seems on the surface to be both basic and imperative is not at all simple and, in fact, is not achievable in the absolute sense. It is extremely hard to accept even the idea that food is relatively—not absolutely—safe. What appears to threaten food threatens in a very direct and visceral way.

An understanding of basic definitions about safety and toxicity is crucial. First of all, compounds, no matter how salutary, can be ingested in some manner or in some quantity that will cause toxicity. Toxicity is the capacity of a substance to produce some adverse effect or harm. Even essential components such as water and vitamins can be consumed at toxic levels. Too much pure water can cause renal shutdown; excessive vitamins can cause minor problems such as flushing or nausea or major problems such as liver damage, teratogenicity, and death. The 1538 Paracelsus motto, "only the dose makes the poison" also operates for food components (Jones, 1992).

Second, what is safe for one is not safe for everyone. Individuals who have allergies, inborn or acquired errors of metabolism, or certain diseases can ingest a food in a usual and customary manner and suffer adverse, even in rare instances, fatal outcomes. Thus, foods that are safe for most are not safe for all.

Third, how the food is used or produced may alter its safety. Combinations of foods and drugs or a certain food with a bizarre or poor diet can render an otherwise safe food as harmful. A food may contain an unexpected contaminant such as a mycotoxin or toxin acquired during certain growing or feeding conditions. Thus, what is usually safe harbors a masquerading toxin.

Since absolute safety is unattainable, relative safety is what is sought. Relative safety is the probability that no harm will come when the food is consumed in a usual and customary manner. Even relative safety, when it comes to food, is a big order because it requires constant diligence by each party who comes in contact with the food. Any glitch in the system from the field to the table can introduce a potential hazard.

The starting material, i.e., the food itself, must not have high levels of naturally occurring toxicants. Safety must be maintained by growing the food in an environment free of pollutants or contaminants. Plant raw materials must be free of infestations, harmful residues, mold, and mycotoxins. Animal raw materials should not contain any veterinary drug residues or any abnormal constituents transferred from feedstuffs. Metal particles, weed contaminants, or other incidental components from harvesting or processing must be vigilantly prevented. During transportation from the field to plant or market, carriers must handle the food to maintain its quality. Care must be exercised so that proper temperatures and moisture levels are maintained and that no infestation occurs during any point of the storage, shipping, and processing. Handling conditions during manufacturing, storage, shipping, and marketing must not allow microbial or chemical contamination. Prevention of further contamination is often done through packaging together with other techniques such as modified atmosphere packaging. Packaging and the other applied techniques must not introduce risks of their own such as micro-migration of nonfood polymers into the foodstuff or alternative microbial risks.

Once in the consumer's hands, food must provide the expected nutrients during its shelf life. During preparation in a home or food service setting, further contamination with microorganisms or other hazards must be prevented. Thus, what is taken for granted as both simple and imperative is anything but.

Furthermore, even if all this happens according to proper protocol, the food may still not be safe because the person ingesting it may react adversely to it, choke on it, or have a food–drug interaction that renders a usually safe food component as injurious. While scientists know that safe is not risk-free, this concept is extremely difficult to convey to consumers.

In addition to the tension created by convincing consumers that no food is risk-free, tension arises from both the fear of what is no longer familiar and the scientific and technological complexities of modern life. Unlike the childhood story where the little red hen grows the wheat, harvests and mills the wheat, and bakes the bread, most modern consumers are far removed from the production and processing of food. Some are even separated from preparation of food. In countries like the United States and the UK, safe storage and preparation techniques are no longer learned at home because very little food is prepared. Plated meals and deli foods that are heated to serving temperature, usually in a microwave, have become the norm for some families.

CONSUMER ATTITUDES TO THE FOOD SAFETY PROBLEM

Consumer concerns about food safety are, in part, a protest to scientific and technological complexity and lack of trust in government, big business, and advertising. For some, science has become the problem, not the answer. Even those with belief in the promise science holds, scientific complexity confuses. Two experts on a topic espouse diametrically opposed positions and many interpretations of the same data. Different assumptions and extrapolations lead to entirely different conclusions. Self-appointed experts and consumer groups with specific agendas add to the multiplicity of positions consumers hear. It is nearly impossible for the consumer to sort through the cacophony to attempt to find the truth. This is made worse by the fact that the pseudo-scientists and self-appointed experts are often louder and easier for the consumer to understand than are scientists because neither of these types is tied to the conflicting evidence produced by the data. Groups that position themselves as anti-big business and -technology and pro-consumer and -environment also have a credibility edge. Exacerbating the problem is the quality of information given to the public. Information about food is often either too simplistic, too boring, too incomplete, or too biased. Sixty-second newsbites drastically distill a 10-year study, take an item out of context, or reflect the findings of a single study that is not in agreement with a whole body of other studies. The news commentator has neither the time nor the knowledge to interpret what this news item means to someone who eats this food once a week. Even worse, there are media reports that take findings and try to scare the public using discredited and highly controversial data. An example of this was the alar scare, which occurred in the United States in 1989. The most egregious data on levels of pesticide in apples were delivered as a press release to a television news magazine by an anti-pesticide environmental group. The report created a major uproar, decreased consumption of apples in the U.S. by nearly one-third, and resulted in news stories in virtually every paper in the nation. National news magazines did features questioning whether anything was safe to eat.

Several problems confound the issue of reliable information. Human behavioral studies found that sources were more believable if they reported that a food or component is not safe, rather than affirming that a food is safe (Occhipinti and Siegal, 1994).

Along with increased complexity, there is, in some countries such as the United States, increased scientific illiteracy coupled with greater fear of chemicals and technology. The realization that human life has always entailed exposure to chemicals and that everything ingested and inhaled is composed of chemicals is not a shared assumption. Even more elusive to most consumers is the fact that naturally occurring chemicals are more abundant and can be more toxic than synthetic ones. Most consumers believe the converse — that natural chemicals are innocuous and synthetic ones are nefarious. Ironically, for food chemicals, much more is known about the synthetic chemicals added to food than those that occur naturally.

TESTS TO DETERMINE FOOD SAFETY

INTRODUCTION

Human beings have always been intuitive toxicologists, relying on their senses of sight, taste, and smell to detect harmful or unsafe food, water, and air. When asked, some consumers still feel that they are the prime determiners of food safety and try to rely on themselves. Scientists and savvy consumers have come to recognize that our senses are not adequate to assess the dangers inherent in exposure to a chemical substance especially one in which the ill effect is either cumulative or delayed. The sciences of toxicology and risk assessment have developed to assess the safety of foods and their constituents.

SPECIFIC TESTS

Toxicity tests are performed on all compounds used as intentional additives and pesticides. Acute tests using at least two species of experimental animals determine lethal dose 50 (LD_{50}) — the dose that kills half of the animals.

Metabolic tests are done early in the protocol to track the fate of the compound in the body. If metabolites are formed, their fates and toxicity must also be determined. Different species are used to test whether the metabolism is the same and to see which species will be most similar to humans.

Subacute tests are then performed that feed an array of doses below the LD_{50} to at least two species for two to three months. A threshold or no observable effect level (NOEL) is determined from the highest dose that produces no harm in the most sensitive species.

CHRONIC TESTS

Feeding a compound at doses 1,000 to 100 times what a human would likely ingest determines chronic toxicity. Two to three species fed for a lifetime are used in these tests, which not only assess the health of the animal, but determine if there are any reproductive or offspring abnormalities.

POST-TESTING

After the various forms of testing are done and the various tests indicate that an additive may be safely used, the NOEL is divided by a safety factor so that the acceptable daily intake (ADI) can be determined (Figure 11.1). The ADI is expressed as milligrams of the test substance per kilogram of body weight per day. The safety factor is arbitrary and may vary according to the test material and circumstances. Often, a factor of 100 is chosen. The rationale for 100 is that, if the average sensitivity of humans to a particular compound is ten times greater than for that of the most sensitive test animal and if the most sensitive humans are ten times more sensitive than the test animal, the use of the factor $10 \times 10 = 100$ would mean that the most sensitive individual could safely ingest the amount equivalent to the ADI (Francis, 1993).

Carcinogens may require special consideration in choosing a safety factor. The 100-fold safety factor may be inadequate, and factors as high as 5,000 have been proposed. There is fear that for some carcinogens there may be no threshold or tolerance level. Deciding the question for carcinogens is very risky because there are several theories about carcinogenicity

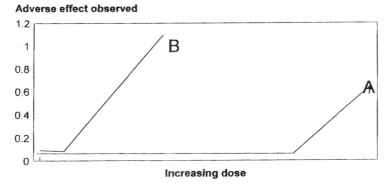

Figure 11.1 No observable effect level. A: Chemical with high no-observable effect level; B: Chemical with low no-observable effect level.

of chemicals at low levels (Francis, 1986; Ames and Gold, 1990). Decisions are difficult because the lag time between exposure to a carcinogen and the development of a tumor may be at least 20 years, and various other aspects may favor promotion of detoxification.

For all chemicals, carcinogens or not, a final step must be completed after establishing the ADI; a value must be assigned for the amount that would be allowed in a particular foodstuff. Consumption estimates for foods or commodities that might contain the chemical are needed along with estimates of total intake from all sources. This calculation is used to establish the maximum residue level (MRL) for any particular commodity.

Constant monitoring and reevaluation of these estimated intakes make certain that estimates reflect real exposure. Recent data indicate that the intake estimates used to calculate the MRL may need some modification. More specific food product consumption data for high-risk groups are needed. Sampling procedures are often too aggregate to target these groups. More consideration of food consumed away from home is needed because this market segment now accounts for about half of all U.S. consumer food expenditures.

Even with the best scientific methods, extrapolations and judgments are required in order to infer human health risks from animal data. Basic toxicological concepts, assumptions, and interpretations were found, according to a recent survey, to differ greatly between toxicologists and laypeople, as well as differences among toxicologists working in industry, academia, and government. Toxicologists were found to be sharply divided in their opinions about the ability to predict a chemical's effect on human health on the basis of animal studies (Neil et al., 1994).

FOOD SAFETY CONCERNS

The top food safety concern for EC, Australian, and U.S. government officials is foodborne disease. Consumers now perceive it as a great concern but have greater fear about food grown with pesticides. Highly publicized outbreaks of *Salmonella, Listeria,* and *E. coli* have raised consumer concern so that fear of foodborne disease is nearly equal to that of fear of chemicals.

The scientific groups judge a hazard on known deaths or cases of illness, and consumers still have many concerns about numerous chemicals in the food. With science increasing its ability to detect substances, now to the attogram level, more must be done to help the consumer move from the position of "zero tolerance."

Pathogenic bacteria are responsible for the majority of food-related outbreaks in the United States. Opportunity for contamination exists at every stage in the food chain. The actual incidence of foodborne disease is unknown, even in countries with fairly sophisticated monitoring systems because the number of cases are severely underreported. For instance, in the United States, there may be as many as 5 million outbreaks per year and 81 million cases of diarrhea due to food. Furthermore, there is some evidence to support that chronic diseases such as arthritis may be triggered by foodborne bacteria (Archer, 1988). Groups most susceptible to the ill effects of contamination are the very young, the very old, the chronically ill, and the immunocompromised. *Salmonella* species, *Staphylococcus aureus, Clostridium perfringens, Campylobacter jejuni, Yersinia enterocolitica,* and pathogenic strains of *E. coli* are responsible for much of the foodborne disease. *Listeria* and botulism are also of concern because of the high degree of morbidity and mortality. The increased incidence in all types of viral, bacterial, and parasitic infections is not only due to better reporting, detection, and surveillance, but also due to changes in our consumption patterns and greater pollution, as in the case of *Vibrios*. People in many Western nations buy more preprepared, prepackaged foods; demand out-of-season and exotic foods from all around the globe; utilize new technologies such as modified atmosphere; demand food with less salt and fat; and use food service and delis more often. All these trends can impact the microorganisms and chemicals found in food. A study by the U.S. Centers for Disease Control traced 77% of the microbial disease outbreaks to food service establishments, 20% to homes, and 3% to food processing plants. From these data, it can be seen that while increased food safety programs such as HACCP (Hazard Analysis Critical Control Point) or LISA (Longitudinally Integrated Safety Assurance) are being mandated in many parts of the world, there needs to be a plan to reduce problems in food service and an education program to tell consumers about safe food handling and preparation.

Other things being equal, consumers prefer more, rather than less, food safety. The preference for food safety must be balanced against other attributes of the product. A consumer's preference for a soft cheese made with unpasteurized milk may outweigh the small risk of contracting *Listeria* or *E. coli* from that source. Price may impact the choice of a specific food, if the safer food is more expensive to produce. Interestingly, food safety is an income elastic good. As incomes rise, so does the sizeable premium a consumer will pay for food that is perceived to be safer (Swinbank, 1993).

The market mechanism does not work directly with respect to food safety issues. First, food safety attributes are not readily determined at the point of purchase. Competing food products are rarely appraised according to alternative levels of microbiological spoilage or probability of contaminated cans. Second, a less safe product is usually not intentionally supplied. The problem often lies in accidental contamination or a change in the process or product such that microbial growth is allowed or contamination occurs.

Consumers react rather violently to a dread outcome. If a supplier makes a mistake, the company will be pilloried in all forms of the media. One mistake in one container in a million (an undetectable number) that gives rise to the death of two people will not only affect the sale of the particular product, but also of all products of that type and by that manufacturer. Many companies fold in the face of such an adverse event.

To prevent such negative consequences, many industry programs are in place worldwide. HACCP, good manufacturing practice (GMP), and other such programs require sanitary conditions for food production and help to eliminate chances for contamination. ISO 9000 and ingredient specification programs ensure that the raw materials incorporated into food products meet required low levels of contaminations and reduce both corporation and consumer risk.

RISK/BENEFIT AS IT APPLIES TO FOOD

PERCEIVING OF RISK AND BENEFIT

The risk/benefit concept is clear-cut in many aspects of our lives. For instance, in medicine, the treatment of disease has inherent and sometimes lethal risks, but the benefits afforded by the treatment are believed by most to outweigh the risks. Here, the risk is vital and the benefit is vital. The health benefits of exercise and exhilaration from skiing, for many, outweigh risks incurred when skiing. Here, the risk and benefit are nonvital, but voluntary. With food, the risk/benefit may be much less apparent to the consumer. Consumers frequently are barraged by the risks of using pesticides and the possible risks of pesticide residues in food. The risks range from possible increases in the number of cancers to reduction in the immune response and environmental concerns. The benefit piece is rarely part of the news coverage. Articles stating that pesticides reduce vector-borne disease, decrease the amount of fossil fuel required to mechanically cultivate a field, reduce the number of bug parts and droppings in food, and increase crop yield to feed a burgeoning global population are rarely cited in stories about pesticides.

In the same manner, risks of additives are often cited by various groups, while the benefits remain unsung in articles to which the consumer has ready access. For instance, a preservative can have several important benefits. One, formation of a cancer-producing mycotoxin may be arrested by inhibiting mold growth. Two, food costs can be reduced because staling and oxidation are retarded. Three, food waste is reduced. Four, oxidized health products with their attendant health risks are reduced. Five, preservatives make possible foods that meet consumer wants of convenience or special needs such as low salt.

Looking at the risks, and not the benefits, of chemicals or technologies associated with food is like an accountant shearing a ledger in half and considering the liabilities without the assets. However, even if judgments of risk and judgments of benefit are considered, they have been found to be inversely related (Alhakami and Slovic, 1994). Activities or technologies that are judged high in risk tend to be judged low in benefit and vice versa. In situations where there is high risk, there tends to be a confounding of risk and benefit in people's minds. Risk gurus Paul Slovic and his colleagues at Decision Research in Eugene, Oregon found that perceived risk is often not related to the probability of injury.

COMMUNICATING RISK

Communicating risk is about trust. Trust can be easily destroyed but is very difficult to establish. Once there is the element of distrust, it fuels more distrust. Interestingly, sources of bad news are more credible than sources of good news.

Risk communicators should be aware that risks that people can choose to avoid—such as skiing—that give a sense of control, that are familiar—such as *Salmonella* from potato salad or disease from smoking—and that have been around a long time—like foodborne disease—are easily tolerated and often minimized. Risks that are poorly tolerated are involuntary, such as exposure to small amounts of pesticide residues in food; have unknown effects such as biotechnology-produced tomatoes; or have long delayed effects such as are possible with pesticide exposure.

The framing of risks makes a tremendous difference in their acceptance. A study reported in the *New England Journal of Medicine* showed that 44% of the respondents would select a procedure for a lung cancer treatment when told they had a 68% chance of surviving, but only 18% would select the procedure if they were told they had a 32% chance of dying. It is no wonder that the consuming public has an inordinate fear of some food additives and pesticides because the information is always reported in terms of increased risk, rather than possible reduced risk due to less mold or bacterial growth.

Some risks were accepted when consumers were faced with a choice about risks. For instance, consumers in California were asked if they would accept food irradiation. The responses were varied but had strong negative leanings. If the same consumers were asked if they would accept irradiation as a way to use less pesticide or reduce a microbial hazard, then irradiation acceptance increased (Bruhn, 1995).

Consumers, thus, appear quite capable of understanding and using the

risk/benefit concept for making decisions about food choices. It is mandatory that all sides of the story get a hearing so that these informed choices
can be made considering both the risks and benefits.

The final word about risk/benefit is that the definition of the risks and the
benefits are not the same in all cases. The definition of benefit needs to be
clearly defined. For some countries, the benefit is reduced postharvest
food losses. The value of this benefit can vary with the country involved.
Different weights may be assigned if there is danger of severe food short-
age versus simply an increase in the cost of the substance. Thus, the
science and art of risk assessment are difficult because each step in the
process is value-laden.

NUTRITIONAL EVALUATION OF FOOD PROCESSING

INTRODUCTION

Food processing both maintains and destroys nutrients. In a few cases,
nutrients are even more available after processing. This seeming paradox
will be explored.

For most foods, processing reduces the nutrient content when compared to the freshly harvested item. However, processing makes food
and the nutrients within available at a later date so the loss incurred
must be compared with what would happen to the diet if no processing were utilized. For some foods, processing the food makes the
nutrients more available because toxic factors and antinutrients are
destroyed. Cassava, soybeans, and corn are all examples of important
classes of food made either less toxic or more nourishing through processing. Cyanide is removed from cassava with grinding and soaking, enzyme inhibitors and lectins are destroyed during the heating of the soybean, and niacytin releases niacin when corn is processed with lime
(CaO).

All nutrients (except water) undergo either chemical or physical changes
during processing that render them inactive or less bioavailable. This
occurs for macro- and micronutrients. During browning, the amino acid
lysine can react with carbohydrates to lower the biological value of protein. Fat oxidation can decrease the level of essential fatty acids in the diet
and can lower the overall food quality by introducing radicals and other
oxidized products into the diet. Altering particle size of grains is one way
that processing can change the physiological effect of nondigestible carbohydrates. Pregelatinizing starch causes greater elevation of blood glucose
than the equivalent starch item, which has not undergone the pregelatiniza-
tion treatment.

EFFECTS ON VITAMINS

Vitamin loss through different processing procedures and in various foodstuffs has been studied. The most labile of the vitamins is vitamin C. Vitamin C is lost during canning, blanching, frozen storage, drying, and irradiation. Losses can be 100% from a food if the conditions of processing and storage are not controlled.

A recent study of processed mashed potatoes fortified with vitamin C gives evidence of the lability of this vitamin. Cumulative losses of vitamin C were 56% for adding the vitamin to freshly mashed potatoes, 82% for drum-dried potatoes, 82% for flakes stored 4.3 months at 25°C, and 96% reconstituting mashed potatoes and holding them 30 min on a steam table. One serving (100 g) would contain 10 ppm—2% of the adult RDA (USA). A more stable isomer used to fortify the mashed potatoes would yield about 201 ppm (about 33% of the RDA of vitamin C per serving) (Wang et al., 1992).

Thiamin and folate, while not as labile as ascorbate, are easily lost during processing. Thiamin is extremely soluble and destroyed by heat, so much can be leached into the cooking or storing liquids during preparation of both meats and vegetables. Losses in the making of soy flour are minimal, but losses in the making of soy flour into tofu stored in water are 85% or greater (Fernando and Murphy, 1990). In addition to losses due to leaching, thiamin content can decrease markedly when subjected to basic pH. Alkalinity causes the two rings of thiamin to split, which causes loss of all biological activity for humans. A quick bread with bicarbonate leavenings can lose up to 75% of the thiamin in the finished baked product because of the synergistic effect of both heat and pH. Sulfites used as preservatives will also cleave thiamine's two rings and will thus destroy thiamine.

Folic acid is easily lost during storage of fresh vegetables at room temperature and through many heat processes. Oxidative destruction of 50–95% of the folate can occur with protracted cooking or canning.

Riboflavin is unstable to light and thus riboflavin-containing foods subjected to either ultraviolet or visible light can show significant losses of riboflavin. Riboflavin is much less soluble than thiamin, but long-term storage in water can cause leaching. For instance, tofu stored in water can lose 80–90% of the riboflavin.

Pyridoxine (vitamin B-6) losses in food are dependent on the temperature and on the specific form of the vitamin. Thermal processing and low-moisture storage of certain foods results in reductive binding of pyridoxal to lysine in proteins making it unavailable. Thus, canning and drying losses can be substantial. Losses in canned infant formulae are of particular concern because the formulae may be an infant's only food source. During blanching, there is not much measurable loss of vitamin B-6 content, but recent studies have shown that loss of bioavailability or absorbability may be significant. Most meats, a good source of vitamin B-6, luckily lose little of it during preparation.

Fat-soluble vitamins show somewhat greater stability than water-soluble vitamins. They are not as easily destroyed by normal cooking and are not leached into the cooking water. However, vitamin A changes slowly from the all-*trans* form, the most biologically active form, to the *cis* form during canning and with long times on a steam table. Carotenoids, currently valued for the antioxidant and possible anticarcinogenic potential, also oxidizes to some degree during heat treatment. Traditional canning causes greater losses of vitamin A and carotenes than does high-temperature, short-time (HTST) processing (Chen et al., 1995). Little loss occurs in frozen blanched vegetables, but vitamin A and other carotenoids easily oxidize in drying when no anti-browning additives such as sulfite are used.

Vitamin E is not very stable to heating followed by freezing and is lost in milling as the germ is removed from white flour. Some vitamin E isomers are lost during the processing of oil. In many countries, the vitamin D content of food is increased through fortification of dairy products. Vitamin K is not greatly affected by heat but is lost to light, so vitamin K-containing oils retain their vitamin content if stored in amber bottles.

EFFECTS ON MINERALS

Minerals are lost into the cooking liquid if the liquid is not ingested. Minerals are retained in the bran and germ fragments of the grain and therefore lost to those who ingest only refined grain products. Some minerals are released and made more available during the cooking process while others become less bioavailable. While effects of various nutrients and certain nonnutrient components of food on mineral utilization have been intensively studied, less is known about the effects of food processing and preparation procedures. Fermentation during the production of beer, wine, yogurt, and African tribal foods affects bioavailability of zinc and iron. Baking determines the chemical form of iron in fortified bread products, and these changes can affect bioavailability. Availability of iron in milk-based infant formula depends on whether iron is added before or after heat processing. Food packaging (e.g., tin cans) can alter food composition and thus potentially affect mineral bioavailability. Maillard browning has been reported to cause slight decreases in zinc availability.

NEWER AND NOVEL TECHNOLOGIES

Treating fresh or frozen meats with ionizing radiation is an effective method to reduce or eliminate foodborne human pathogens. Irradiation dose, processing temperature, and packaging conditions strongly influence the results of irradiation treatments on both microbiological and nutritional quality of meat. Radiation doses up to 3.0 kGy have little effect on the vitamins in chicken or pork but have very substantial effects of foodborne pathogens. Even vitamins, such as thiamin, which are very sensitive to ionizing radiation are not significantly affected by the U.S. Food and Drug Administration maximum approved radiation dose to control *Trichinella*, but at larger doses it is significantly affected (Fox et al., 1995).

Another process technology is both new and old. Using fermentation, plant and animal husbandry and other biological processes are employing biotechnology. However in the last twenty years, the technology has taken a giant step forward in the form of genetic engineering and can use gene splicing to both speed up and make the process more precise. It also can allow changes that were never conceivable before. Biotechnology has the potential to both increase and decrease available nutrients in the same way that plant breeding can. Care must be exercised that the nutrient content of foods is not reduced when another positive attribute is engineered into the food.

One food safety concern about genetic engineering is that toxic components naturally found might be increased. Attempts to breed or genetically engineer plants with natural herbicides or pesticides or herbicide-resistant plants could increase the potential toxicity of the foodstuff. This must be carefully monitored. Scientists must not be lured by the trap that nature is benign and chemicals from the lab are noxious.

ADDITIVES

Food additives can enhance the safety and nutritional quality of a food or vice versa. By preventing oxidation of fat and easily oxidized vitamins by antioxidants, safety is enhanced and the intended nutritional value of the food is delivered. Anti-browning agents such as sulfite retain vitamins A and C while lowering the amount of thiamine, folate, and pyridoxal. Sorbic acid can prevent mold and possible mycotoxins but can form protein adducts in the stomach, which affect the availability of the protein. Nitrate interacts with vitamins C and E to prevent the formation of nitrosamines. This reaction will use the vitamins, and thus, they will not be available for other functions. Phosphates are antimicrobial because of their sequestering capacity. This capacity can have an antinutritional effect because of mineral-binding ability. Vitamins themselves are added to fortify and enrich products, making the food have more nutrients than it might otherwise. Sometimes food is overfortified, giving consumers the idea if they eat one serving of highly fortified food, they needn't pay attention to other parts of the diet.

New products that replace entire foods or macro-nutrients, such as fat replacers, must be evaluated both for their nutritional contribution and for their dietary impact. These foods can do wonders in helping certain people reach needed dietary goals. In other cases, they introduce risks such as overconsumption and abuse, vitamin leaching or competition, and gastrointestinal problems. In some instances, additives that were intended for use in micro amounts are now being used in macro quantities. They may exceed the ADI and have effects that would make them unsafe at high levels of use.

Additives, like all food components, need to be looked at with the risk and benefit in mind. Contrary to popular belief, food additives can be regarded as the safest and most studied constituents of our food supply. Yet this is as it should be. Furthermore, surveillance of food additives is a mandatory consequence of their use, in that their safety must be continually assured considering any change in usage patterns and new methodological advances. This constant vigilance makes consumers feel that science and technology is untrustworthy because the answers seem to change. In fact, just the reverse should be true, because additive safety is constantly being challenged to ensure that only the most wholesome food products are on the market. Substances should never be added to food without careful analysis and a conservative approach to their use. A clear benefit to the end user must be present before a food additive should be allowed.

REFERENCES

Alhakami, A. S. and Slovic, P. 1994. "A Psychological Study of the Inverse Relationship between Perceived Risk and Perceived Benefit," *Risk Analysis,* 14(6):1085–1096.

Ames, B. N. and Gold, L. S. 1990. "Too Many Rodent Carcinogens: Mitogenesis Increases Mutagenesis," *Science,* 249:970–971.

Archer, D. L. 1988. "The True Impact of Foodborne Infections," *Food Technology,* 42(7):53–58.

Bruhn, C. M. 1995. "Strategies for Communication Facts on Food Irradiation to Consumers," *J. Food Protection,* 58(2):213–216.

Chen, B. H., Peng, H. Y., and Chen, H. E. 1995. "Changes of Carotenoids, Color, and Vitamin A Contents during Processing of Carrot Juice," *J. Agricultural and Food Chemistry,* 43(7):1912–1918.

Fernando, S. M. and Murphy, P. A. 1990. "HPLC Determination of Thiamin and Riboflavin in Soybeans and Tofu," *J. Agriculture and Food Chemistry,* 38(1):163–167.

Fox, J. B., Lakritz, L., Hampson, J., Richardson, R., Ward, K., and Thayer, D. W. 1995. "Gamma Irradiation Effects on Thiamin and Riboflavin in Beef, Lamb, and Turkey," *J. Food Science,* 60(3):596–598, 603.

Francis, F. J. 1986. "Testing for Carcinogens." *Science of Food and Agriculture (CAST),* 4(2):13–15.

Francis, F. J. 1993. "How Do We Test for Safety of Food?" *Science of Food and Agriculture,* 5(1):2–5.

Jones, J. M. 1992. *Food Safety.* Eagan, MN: Eagan Press.

Neil, N., Malmfors, T. and Slovic, P. 1994. "Intuitive Toxicology: Expert and Lay Judgments of Chemical Risks," *Toxicology and Pathology,* 22(2):198–201.

Occhipinti, S. and Siegal, M. 1994. "Reasoning about Food and Contamination," *J. Personality and Social Psychology,* 66(2):243–253.

Swinbank, A. 1993. "The Economics of Food Safety," *Food Policy,* 18(2):83–94.

Wang, X. Y., Kozempel, M. G., Hicks, K. B., and Seib, P. A. 1992. "Vitamin C Stability during Preparation and Storage of Potato Flakes and Reconstituted Mashed Potatoes," *J. Food Science,* 57(5):1136–1139.

CHAPTER 12

Mutagenic and Carcinogenic Components in Foods

AGNIESZKA BARTOSZEK

INTRODUCTION

THE factors and substances able to induce changes in the genetic code are called mutagens. Those that can cause cancer, excluding genetic susceptibility, are called carcinogens. Such factors are omnipresent in the human environment; they can be of natural origin or can be formed as a result of numerous chemical processes. To these factors belong a variety of synthetic chemicals, combustion products, water and air pollutants, sunlight and ionizing radiation, cigarette smoke, alcohol, and some food components. At the same time, many kinds of foods contain substances displaying anticarcinogenic properties.

It has been estimated that approximately 35% of cancer deaths in the U.S. are attributable to dietary habits. When, in these estimations, the largely voluntary exposure to tobacco is ignored (ca. 30% of cancer deaths), this fraction increases to about 50% (A Scientific Status Summary, 1993). Therefore, it is not surprising that the presence of potential mutagens and carcinogens, as well as anticarcinogenic substances in foods, has become of widespread interest.

Mutagens and carcinogens found in food products can be classified into three categories (Sugimura and Sato, 1983). The first category includes natural compounds such as mycotoxins and substances of plant origin. The second category contains substances formed as a result of food storage, cooking, and processing. The third category of foodborne mutagens and carcinogens is derived from pesticides, fungicides, and artificial additives. Contrary to public beliefs, compounds belonging to the third group appear to have only marginal significance in etiology of human cancer due to very low concentrations to which people are exposed.

The majority of mutagens and carcinogens found in foods is formed during food processing, especially thermal processing. However, processing and heating foods have invaluable advantages. They increase shelf life of foods, which can then be economically priced; decrease the risk of diseases caused by foodborne pathogens; improve taste and nutritive value of food; and provide easy-to-prepare and time-saving convenience foods. Therefore, it is of utmost importance to establish the processes responsible for mutagen and carcinogen formation in food and to clarify the involvement of such substances in transformation of a normal cell into a cancerous one. This field of research is progressing rapidly, and it may be expected that the gathered knowledge will bring about new technologies that will combine current benefits of food processing with minimizing the formation of harmful compounds.

THE ROLE OF MUTAGENS IN CARCINOGENESIS

Transformation of a normal cell into a cancerous one manifests itself macroscopically as an uncontrolled cellular growth, resulting in the formation of a tumor, which may then metastasize and eventually lead to death of the organism. At the onset of cancer development, two major stages can be distinguished: initiation and promotion. Carcinogens responsible for the initiation of the changes that can lead to the conversion of a healthy cell into a neoplastic cell, hence acting at the initiation stage, are divided into genotoxic and epigenetic (Taylor, 1982). Epigenetic mechanisms are poorly understood but are thought to include cancers induced by hormonal imbalances, irritant solids or particles (e.g., asbestos), and tissue damage due to, e.g., repeated injections. Genotoxic carcinogens mainly include substances displaying mutagenic properties. Most mutagenic and carcinogenic food components belong to the group of genotoxins. The name originates from their ability to damage cellular genetic material. The damage usually involves the formation of a covalent bond between an active form of a mutagen and DNA. The sites that most frequently undergo such modifications are nitrogenous bases, guanine in particular (Swenberg et al., 1985). The presence of modified nucleotide, which is usually called a DNA adduct, interferes with proper pairing between complementary bases (adenine and tymine or guanine and cytosine), making the correct DNA replication difficult. The possibility of misreading the information about sequence of bases increases the probability of incorporation of incorrect nucleotide into the daughter DNA molecule. In this way, as a result of replication, the promutagenic lesion, such as DNA adduct, unless repaired by cellular DNA repair systems, becomes fixed in a form of mutation, that is as a change of the genetic code. The altered DNA sequence in turn gives rise to the synthesis of a protein with altered amino acid sequence. Mutation may have no impact on the cell; however, it becomes essential when the biological activity of a crucial protein is impaired.

Carcinogens can initiate the process of neoplastic transformation in the previously described way; that is, they can make the cell become susceptible to cancerous growth. This growth will, however, not take place unless the cell is exposed to one or more factors called promoters. Animal studies suggest that promoters do not possess carcinogenic properties, but they increase frequency of tumor formation and shorten the time necessary for tumor development following the administration of a carcinogenic compound. Promoters found in food include, among others, polyunsaturated fatty acids, e.g., linoleic acid (Hecht, 1993), potassium bromate used in flour processing, polychlorinated biphenyls present in chlorinated drinking water, and even high fat, calorie, or protein intake (A Scientific Status Summary, 1993). Mechanisms of promotion are less well understood as those of genotoxin action, but they are thought to involve stimulation of cell proliferation, blockage of communication pathways between normal and mutated cells, oxidative DNA damage such as hydroxylation of bases, and others (Preston-Martin et al., 1990).

The accumulating evidence indicates that human cancers do not result from one single mutation, but develop in stages reflecting many genetic events. The risk of contracting this disease increases in humans with age to the fourth or fifth power, which suggests that neoplastic cell transformation requires four or more genetic changes (A Scientific Status Summary, 1993). Currently, much research is carried out to verify the hypothesis that cancer development is an aftereffect of the activation of transforming genes, so-called oncogenes, by carcinogenic factors. This activation would consist of one or more mutations of normal genes, named by analogy proto-oncogenes, that code for proteins taking part in the regulation of cellular growth and division as well as signal transduction between cells. The product of the mutated proto-oncogene does not function properly, and the control of cell divisions is impaired. In certain cases, the cell whose genome has undergone the above changes would acquire the capability of uncontrolled proliferation giving rise to a clone of transformed daughter cells – to cancer. At least some carcinogens present in food products are able to initiate such a sequence of events in higher and lower vertebrates (A Scientific Status Summary, 1993).

METABOLIC ACTIVATION AND FORMATION OF DNA ADDUCTS BY FOOD MUTAGENS AND CARCINOGENS

The majority of genotoxic substances found in food do not possess mutagenic and carcinogenic properties in themselves. In order for these properties to be revealed, the metabolic activation in an organism is required, which leads to the formation of electrophilic metabolites capable of binding with nucleophilic centers in DNA. Therefore, in the literature, the term *promutagen,* or *procarcinogen,* is often used to describe the compounds that must be converted by cellular enzymes into ultimate mutagens and carcinogens.

Many enzymatic systems are involved in metabolic activation of carcinogens. The most important is cytochrome P450 complex consisting of at least ten isoenzymes, particularly active in liver. Other enzymes include peroxidases, quinone reductases, epoxide hydrolases, sulfotransferases, and others. Their variety reflects the diversity of chemical structures of compounds to which an organism is exposed. These may be harmful substances as well as needed ones or even those indispensable for its proper functioning. One could argue that the activation of carcinogens is an undesirable side effect of metabolic pathways, which were developed in the course of evolution most probably in order to improve the utilization of nutrients and elimination of unwanted or harmful substances.

Competing with enzymatic activation are detoxification processes that may convert potential genotoxins to less carcinogenic, more readily excreted polar metabolites. To the enzymes responsible for removal of mutagens and carcinogens belong most of all glutathione transferases and glucuronyltransferases; however, activating enzymes are also sometimes involved in detoxification. It even happens that activation and detoxification run in parallel and are catalyzed by the same enzymatic system. For instance, epoxidation of benzo[a]pyrene by cytochrome P450 in position 7,8 results in the formation of a carcinogenic metabolite, while in position 4,5 it produces an inactive derivative excreted readily from the organism. Below are given some examples of well established metabolic activation pathways for a few classes of mutagenic compounds found in food along with the major products of reaction of their main toxic metabolites with DNA, more precisely with guanine, which is the preferred site of binding of electrophilic intermediates.

Metabolic activation of aflatoxin B_1 belonging to the class of mycotoxins is catalyzed by cytochrome P450. The metabolic conversion of this compound can follow many pathways; however, only the epoxidation in position 8,9 produces the ultimate carcinogen:

This metabolite binds to the N^7 position of guanine, giving an unstable adduct 8,9-dihydro-8-(N^7-guanyl)-9-hydroxy-aflatoxin B_1, which either undergoes spontaneous depurination or rearrangement to a stable 8,9-dihydro-8-(2,6-diamino-4-oxo-3,4-dihydropyrimid-5-yl-formamide)-9-hydroxy-aflatoxin B_1 following the opening of an imidazole ring (Wakabayaski et al., 1991):

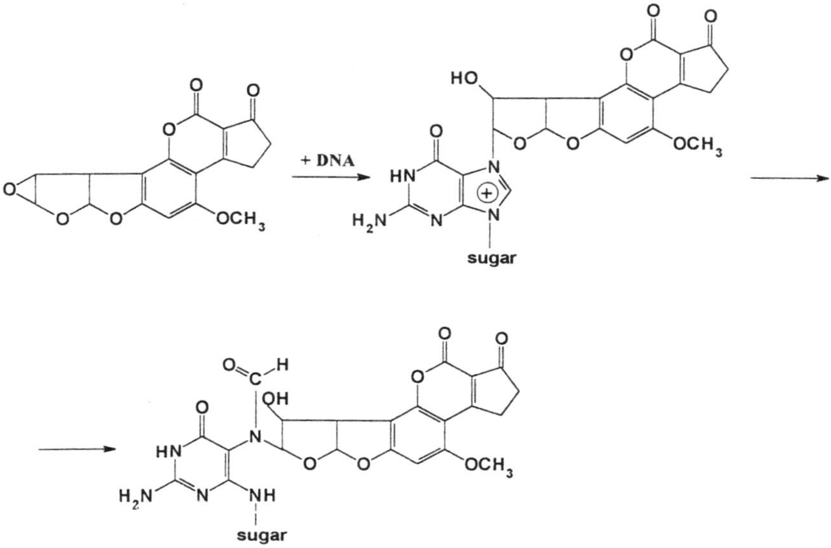

The formation of nitrosamines in the reaction of amines with nitrites under acidic conditions in the stomach can be considered as nonenzymatic activation of amines present in food. Nitrosamines undergo further metabolism, catalyzed enzymatically by cytochrome P450, involving hydroxylation (A Scientific Status Summary, 1993):

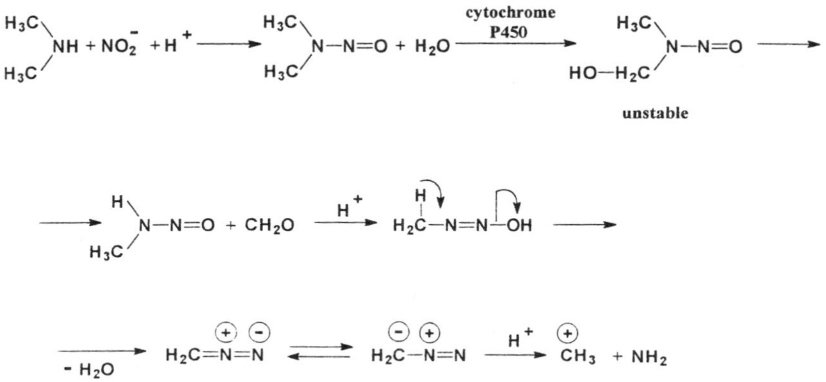

The hydroxylated derivative is unstable and in a series of spontaneous reactions gives rise to methyl carbo-cation, which alkilates guanine in position O^6:

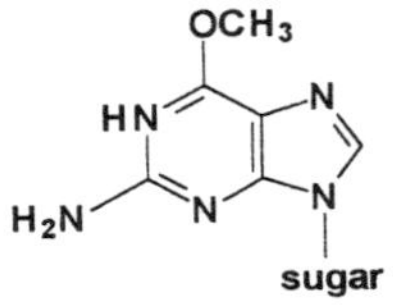

hence, in the site taking part in the formation of hydrogen bonds in DNA with complementary base – cytosine.

Metabolic activation of benzo[a]pyrene consists of three enzymatic reactions:

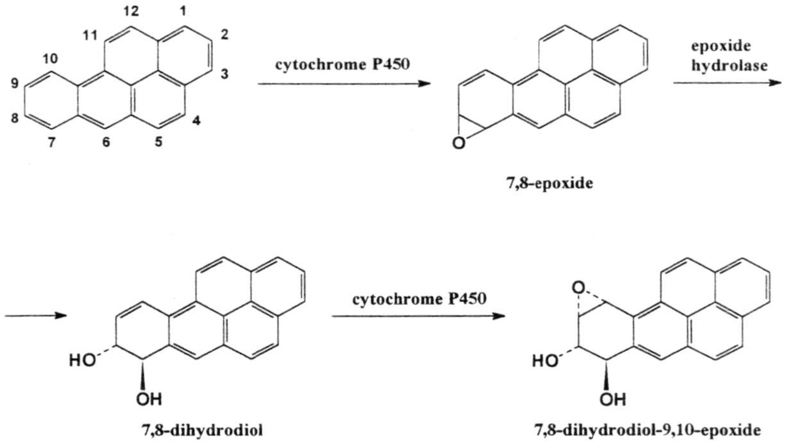

7,8-epoxide

7,8-dihydrodiol   7,8-dihydrodiol-9,10-epoxide

First the formation of epoxide in position 7,8 is catalyzed by cytochrome P450; as has been mentioned before, epoxidation in position 4,5 results in detoxification of this compound. Then, epoxide hydrolase converts the epoxide into 7,8-dihydrodiol, which is subsequently oxidized to 7,8-diol-9,10-epoxide. The formation of four different diastereoisomers is feasible, among which anti-9,10-epoxide derived from (-)-7,8-dihydrol is by far the most carcinogenic (Dipple and Bigger, 1990). In DNA, this derivative reacts most frequently with guanine in such a way that the position 10 of benzo[a]pyrene and N^7 of guanine become linked together.

Aromatic compounds substituted with amino groups, e.g., heterocyclic aromatic amines present in protein food products, are usually activated by cytochrome P450 to hydroxylamines. This type of metabolism is observed in the case of 3-amino-1-methyl-5H-pyrido[4,3-b]indole (Trp-P-2):

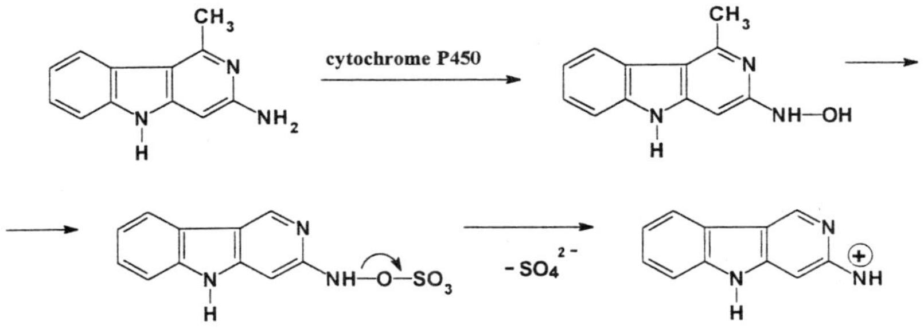

cytochrome P450

unstable

$- SO_4^{2-}$

After further spontaneous rearrangements, hydroxylamine derivatives produce electrophilic intermediates, which are able to modify DNA bases (Sugimura and Sato, 1983). One of the possible structures of DNA adducts formed by Trp-P-2 with guanine is given below:

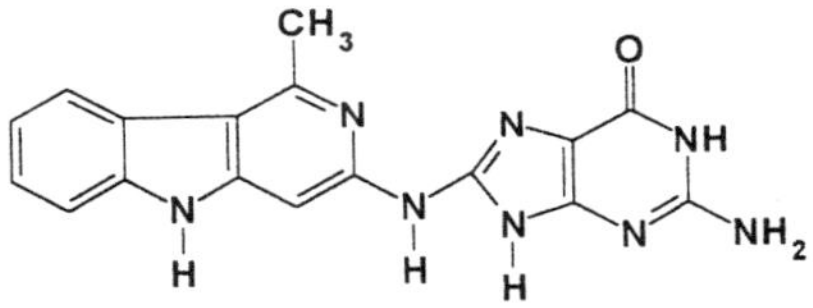

In this case, the position C8 of guanine has been modified. The group of heterocyclic aromatic amines includes so many different compounds that a large variety of chemical structures of DNA adduction products formed by them can be expected.

The enzyme systems implicated in metabolism of carcinogens may be the reason for different susceptibility of humans to cancer. On one hand, the enzymatic activity varies with tissue type, age, and genetic predisposition, and it may be influenced by exposure to inhibitors, inducers, or repressors of a particular enzyme. On the other hand, the potential of carcinogenic compound to damage DNA results not only from its dose, but also depends on the balance between enzymes involved in activation and detoxification of that compound in the individual and target organ.

TESTS FOR MUTAGENICITY AND CARCINOGENIC PROPERTIES OF FOOD COMPONENTS

Food products contain about 2,500 components of nutritive value, numerous ingredients prolonging shelf life of food and also substances formed during processing, pesticide residues, etc. Their safety is of utmost importance for human health protection. In order to evaluate the carcino-genicity of these substances, often of unknown chemical structure, short-term reliable and inexpensive tests are necessary. Since cancer risk asso-ciated with chemical compounds is thought to stem from their ability to induce mutations, mutagenicity is used in the assessment of carcinogenic properties of food components. Such ability can be detected with the aid of bacteria whose culturing is easy, quick, and economical. In the case of bacterial mutagenicity tests, it is assumed that the factors capable of damaging bacterial DNA can interact in a similar way with DNA of higher organisms.

There are many different *in vitro* mutagenicity tests; however, the most widely used is the Ames test (Ames et al., 1975). This test utilizes a mutant strain of bacteria, *Salmonella typhimurium,* unable to synthesize histidine, thus dependent on an outer source of this amino acid. The back mutation in the histidine gene makes bacteria histidine-independent. The frequency of back mutations increases in the presence of mutagenic factors. To make the test bacteria particularly sensitive to mutagens, some strains are additionally mutated to disable their DNA repair systems and increase permeability of the cell wall.

In the Ames test, 10^8–10^9 cells are seeded onto a petri dish containing media with histidine concentration, ensuring only minimal growth of bac-teria. Then a compound or preparation whose mutagenicity is to be tested is applied. Concomitantly with the substance studied, microsomal fraction (usually isolated from rat liver) is also added to mimic metabolic activa-tion of mutagens typical for mammalian cells but often not taking place in bacteria. Only those bacterial cells in which back mutation occurred can grow and form colonies under histidine deficiency conditions. The number of colonies is counted after 48 h of incubation at 37°C. In parallel, the in-cubation of untreated bacteria is carried out to determine the rate of so-called spontaneous mutations. Results are usually recalculated to a number of revertants, i.e., colonies of cells with back mutations, per microgram of the substance tested.

The evaluation of carcinogenicity of substances, that is, their ability to induce cancers, is usually carried out in mice and rats. Several doses of a potential carcinogen are administered to animals. The highest of them cor-responds to the maximum tolerated dose (MTD) that does not cause severe weight loss or other life-threatening signs of toxicity. As a result of such studies, the lowest dose is determined at which carcinogenic effects are still observed. The next level below that is assumed not to have a biologi-cal effect, the so-called "no-effect level." This value, divided by a safety factor of either 100 or 1,000, is considered the acceptable daily intake. The rationale behind the safety factor is that there may be a 10- to 100-fold difference in sensitivity between animals and human beings (A Scientific Status Summary, 1988).

There is a continuous debate regarding whether carcinogenicity of compounds can be indeed predicted based on mutagenicity tests. In other words: are mutagenic substances also carcinogenic? The results obtained with the Ames test seem to confirm the predictive value of such an approach. Among compounds that were mutagenic in this test, 72-91% were also carcinogenic in animal studies. Among those that did not show mutagenic potential, 74-94% did not induce tumors in animals (Taylor, 1982).

From the food safety perspective, it is perhaps worthwhile to mention that, recently, animal carcinogen testing has been challenged by the inventor of the Ames test, who attacked one of the foundations of the U.S. regulatory policy governing potential carcinogens. He has criticized two important aspects of carcinogenicity testing: usage of maximum tolerated doses and performing human cancer risk evaluation in rodents (Ames and Gold, 1990). In the first case, the argument is that, although MTDs do not cause overt signs of toxicity, they may have certain subtle effects, resulting in killing some of the cells and thereby damaging an organ. This will trigger compensatory cell proliferation, hence also DNA replication. The increased rate of DNA synthesis makes fixing of promutagenic DNA lesions in a form of mutation more probable. Some of these mutations may cause cells to become cancerous. The same chemical at lower concentrations, like those to which humans are exposed, might be safe. The second criticism concerns the usage of short-lived species like rodents to estimate the carcinogenic effects in a long-lived species such as the human. In order to achieve a long life-span, humans evolved mechanisms rendering them more resistant to cancers. This makes the extrapolation between the two types of species rather unreliable.

Although neither *in vitro* mutagenicity tests nor carcinogenicity tests in animals can fully reflect the consumers' health risk associated with a given chemical, they play an essential role since they allow one to identify those substances in foods that require detailed toxicological evaluation and whose consumption in larger amounts should not be permitted.

FOODBORNE MUTAGENS AND CARCINOGENS

INTRODUCTION

Epidemiological data and carcinogenesis research over the past 20 years strongly suggest a key role of dietary factors in cancer risk. The majority of mutagens and carcinogens found in food are either natural compounds, mostly of plant origin, or those formed during storage and processing. Mutagens are understood here as compounds giving positive results in the Ames test. Those able to induce tumors in experimental animals are considered to be carcinogenic.

Potential plant mutagens and carcinogens belong to a variety of classes of chemical compounds, e.g., hydrazine derivatives, flavonoids, alkenylbenzenes, pyrrolizidine alkaloids, phenolics, saponins, and many other known and unknown compounds. Plants produce these toxins to protect themselves against fungi, insects, and animal predators. For example, cabbage contains at least forty-nine natural pesticides and their metabolites, few of which were tested for carcinogenicity or mutagenicity, and some of which turned out positive. Some estimates suggest that Americans eat as much as about 1.5 g of natural pesticides per person daily. This is roughly 10,000 times more than the intake of synthetic pesticide residues in food whose intake amounts to about 0.09 mg per person per day (Ames and Gold, 1990). There is currently a heated debate on what risk these natural compounds can pose to humans. Most natural toxins, often found in edible fruits and vegetables, were not tested with regard to their carcinogenic potential as a result of rather nonscientific thinking that only synthetic chemicals are harmful. On the other hand, it is argued that humans evolved defenses to protect themselves against natural toxins, which can be considered a part of human evolutionary history. These mechanisms may or may not be general enough to handle mutagens not occurring naturally in the human environment. Moreover, a diet rich in fruits and vegetables is associated with lower cancer risk (probably owing to the presence of anticarcinogens discussed later in this chapter).

Since the involvement of compounds of plant origin in human cancer risk remains controversial at the moment, it will not be discussed in this chapter. The attention will be drawn mainly to carcinogens and mutagens occurring in foods as a result of storage and processing, because exposure to them, contrary to natural substances, may be, if not totally avoidable, then at least substantially reduced.

MYCOTOXINS

Mycotoxins are highly toxic compounds produced by molds, mostly in the genera *Aspergillus, Penicillium,* and *Fusarium.* They represent the most dangerous contamination arising mainly during storage of numerous food commodities, e.g., corn or peanuts. Tropical and subtropical climates are particularly favorable locations for mycotoxin production because of often poor food harvesting and storage practices.

Among several classes of compounds belonging to the group of mycotoxins, carcinogenic properties have been demonstrated for only three of them. These are aflatoxins and sterigmatocystin inducing liver cancers and ochratoxin A implicated in the development of kidney cancers in experimental animals (Wakabayashi et al., 1991).

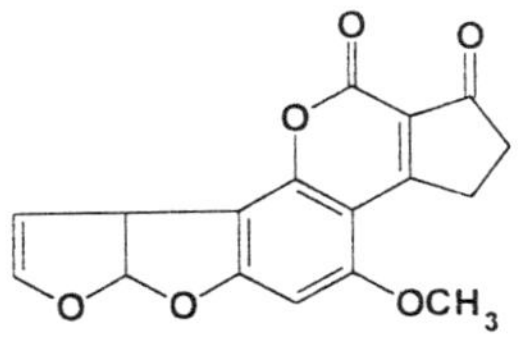

aflatoxin B$_1$

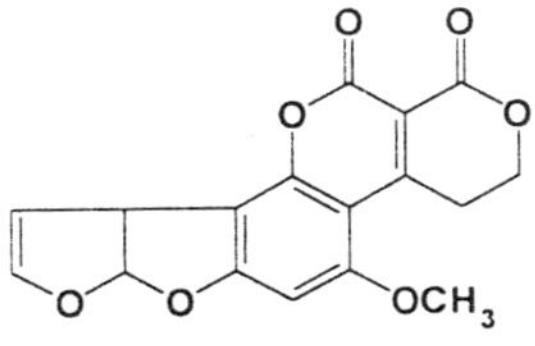

aflatoxin G$_1$

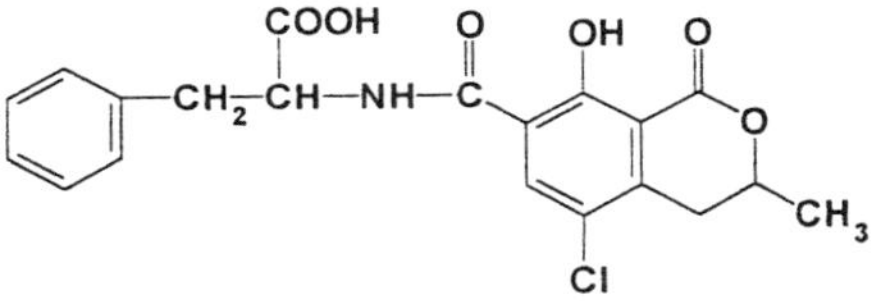

ochratoxin A

Aflatoxin B$_1$ is the most carcinogenic mycotoxin (Table 12.1) and has been classified as a human hepatocarcinogen on the basis of available toxicological and epidemiological data (IARC, 1987).

NITROSAMINES

A number of nitroso compounds, N-nitrosamines among them, are potent carcinogens. Nitrosamines are present in numerous food products. The highest concentrations of these compounds are found in cured meats such as bacon. The formation and types of nitrosamines found in food products have been described elsewhere in this book. Their precursors, especially nitrate, are abundant in some leafy and root vegetables (Table 12.2).

Nitrate and nitrite are also formed endogenously in the human body. In mammalian organisms, following enzymatic conversion of L-arginine, nitric oxide is produced, which in turn, may be converted to nitrite and nitrate (Hibbs et al., 1987). A portion of nitrate, either ingested or endogenously formed, carried out in the blood is secreted by salivary glands into the oral cavity. Here nitrate can be reduced by microbial flora and swallowed. Hence, it ends up in the gastric environment similarly as the nitrite ingested with food. Under the acidic conditions of the stomach, the nitrosation of amines present in food by nitrite occurs, giving rise to N-nitrosamines.

The presence of nitrites has both positive and negative impact on food

TABLE 12.1. **Mutagenicity and Carcinogenicity of Selected Compounds Found in Food.**

Compound	Food Product	Mutagenicity in Strain TA 98 (revertants/μg)	Cancers in Rodents (target tissue/organ)
Aflatoxin B$_1$	Corn	6,000	Liver
N,N-dimethylnitrosamine	Cured meat	0.0	Bladder
Benzo[a]pyrene	Smoked fish	970	Skin
Trp-P-2	Fried meat	104,200	Liver, small and
Glu-P-1	and fish	49,000	large intestine
MeIQ	Meat extracts	661,000	Liver, stomach, lung
Methylglyoxal	Roasted coffee	100[a]	At the site of injection

[a]Data for TA 1000 strain.

safety. On one hand, in many countries, a correlation between stomach cancers, induced probably by nitrosamines, and the amount of nitrites consumed is observed (Fine et al., 1982). On the other hand, nitrites inhibit the growth of *Clostridium botulinum,* thus reducing the risk of food contamination by botulinum toxins. Moreover, under the acidic conditions of the stomach, where they are involved in the formation of carcinogenic nitrosamines, nitrites are capable of neutralizing carcinogens formed as a result of protein pyrolysis (Pariza, 1982).

TABLE 12.2. **The Sources of Foodborne Nitrosamines and Their Precursors: Nitrite and Nitrate (A Scientific Status Summary, 1988; Cassens, 1995; Havery and Fazio, 1985).**

Substance	Source
N-nitrosamines	Cured meat Smoked fish Soy protein foods dried by direct flame Alcoholic beverages Food-contact elastic nettings Rubber baby bottle nipples
Nitrites	Cured meat Baked goods and cereals Vegetables Nitrate reduction *in vivo*
Nitrates	Drinking water Vegetables (beets, celery, lettuce) Nitrate fertilizer residue

Since the presence of nitrites is mainly a consequence of food processing, changes of technology may lead to a considerable decrease of amounts
of these compounds in food products, thereby diminishing the risk of
cancers induced by nitrosamines. Nonetheless, they are likely to remain a
necessary ingredient of preserved foods since an alternative to nitrites as
curing agents and microbiological preservatives has not been found so far.
It has been learned, however, that the formation of carcinogenic
nitrosamines during thermal processing, for example, frying of cured
meats, can be largely inhibited by the addition of antioxidants, e.g., ascorbate and α-tocopherol. The addition of such compounds has now become
a standard procedure (Cassens, 1995).

MUTAGENS IN HEAT-PROCESSED FOODS

Polycyclic aromatic hydrocarbons (PAH) containing a system of condensed aromatic rings are formed as a result of incomplete combustion of
organic matter. Benzo[a]pyrene, a highly mutagenic and exhibiting carcinogenic properties compound (Table 12.1), is the most widely studied
representative of this group. It has been established that for carcinogenicity of these compounds, the metabolites arising from epoxidation of the
so-called "bay region" are responsible.

benz[a]anthracene

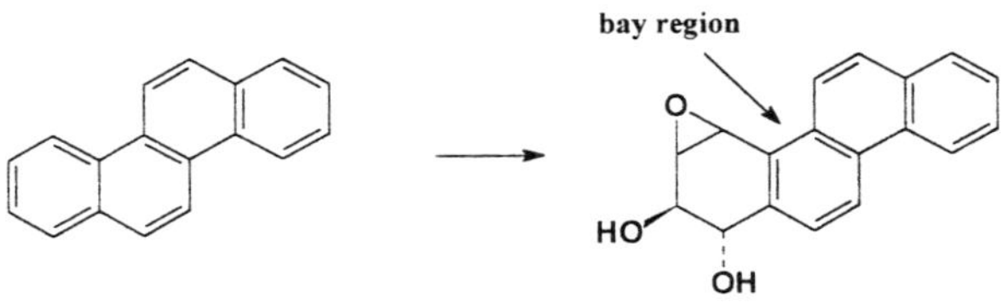

chrysene

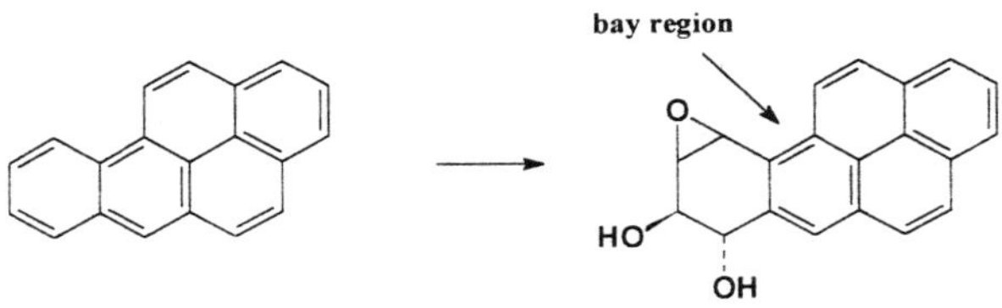

benzo[a]pyrene

In food, PAHs are produced mostly during heating, especially open flame heating, such as grilling of meat. Under such conditions, fat from meat drips onto a hot surface, e.g., hot coal during grilling, and is incinerated. The smoke from the fat pyrolysis containing PAHs adsorbs on the meat. The levels of these compounds that can potentially be produced are relatively large: the surface of a 2-pound "well-done" steak was reported to contain an amount of benzo[a]pyrene equivalent to that found in the smoke from 600 cigarettes (Pariza, 1982). In the case of smoked meat and fish, smoke used during processing is also a source of carcinogenic PAHs (Sikorski, 1988). In addition, a number of food products contain measurable amounts of these hydrocarbons resulting from environment pollution, e.g., fish caught in heavily industrialized regions. The concentrations of PAHs detected in foods are in the range from several to several hundred ng per 1 g of food product (A Scientific Status Summary, 1993).

Heterocyclic aromatic amines (HAAs) are formed during thermal processing of many kinds of foods, especially foods containing much protein. These compounds belong to the strongest foodborne mutagens known (Table 12.1). Studies on animals, including nonhuman primates, demonstrated that HAAs are also carcinogenic, though they usually are not very potent carcinogens (Sugimura and Sato, 1983; Wakabayashi et al., 1991).

The temperature applied during processing has a decisive influence on the kind of HAAs formed in food. The products of amino acid and protein pyrolysis whose chemical structures are given below are produced in temperatures higher than 300°C. Therefore, they are detected mainly in meat and fish subjected to open flame broiling.

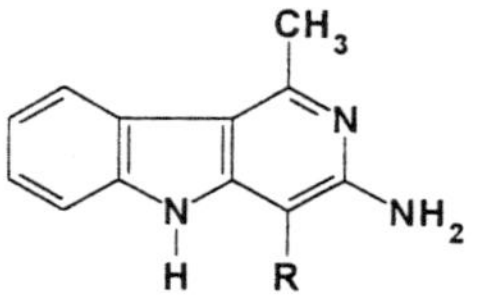

Trp-P-1: R = CH₃   Glu-P-1: R = CH₃   Phe-P-1

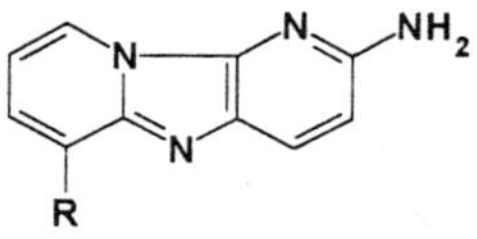

Trp-P-2: R = H   Glu-P-2: R = H

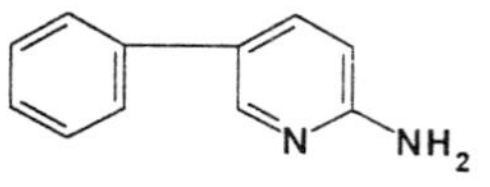

Trp-P-1:   3-amino-1,4-dimethyl-5 *H*-pyrido[4,3-*b*]indole
Trp-P-2:   3-amino-1-methyl-5 *H*-pyrido[4,3-*b*]indole
Glu-P-1:   2-amino-6-methyldipyrido[1,2- *a*:3',2'-*d*]imidazole
Glu-P-2:   2-aminodipyrido[1,2-*a*:3',2'-*d*]imidazole
Phe-P-1:   2-amino-5-phenylpyridine

Mutagens and carcinogens are also found in a wide variety of foods processed under more moderate heating conditions (150–200°C). These are derivatives of quinoline and quinoxaline, formed in the reaction of creatinine with amino acids and sugars.

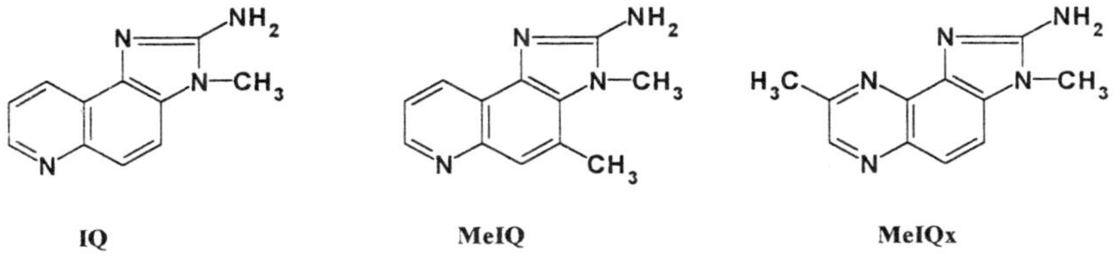

IQ: 2-amino-3methylimidazo[4,5-*f*]quinoline
MeIQ: 2-amino-3,4-dimethylimidazo[4,5-*f*]quinoline
MeIQx: 2-amino-3,8-dimethylimidazo[4,5-*f*]quinoxaline

All the reactants are thus the natural constituents of meat and other products. Since members of this class of HAAs are generated in the range of temperatures to which a product's surface is exposed during frying and baking, they are detected first of all in fried or baked meats, fried fish, bakery products, and also meat extracts (Miller, 1985).

The formation of mutagens in canned foods is also associated with thermal processing, although temperatures applied in this case are relatively low: 110–120°C. Mutagenic substances produced during canning have not been characterized chemically (Krone et al., 1986).

As can be concluded from the above description, common cooking practices, as well as commercial food processing, may lead to the formation of mutagenic substances. Temperature and, to a lesser extent, duration of heating are the major factors responsible for PAHs and HAAs in food. Therefore, human daily exposure to these mutagens varies greatly, depending on dietary habits and cooking practices. For example, a heavily grilled surface and pan scrapings, often used as a base for gravies and sauces, contain 30- to 200-fold greater amounts of HAAs than cooked meat (Gross et al., 1993). Mutagens of this kind were not detected in beef either processed in a microwave oven or stir-fried for 3 minutes on high heat (Miller, 1985).

Although mutagens formed as a result of food heating are often highly-mutagenic, they do not seem to pose a substantial risk to humans since they are not very potent carcinogens and their concentrations in foods are usually very low, in the range of 1 ng per 1 g (A Scientific Status Summary, 1993). Perhaps more worrying are recent reports showing that such food-borne mutagens like those from HAA class are excreted rapidly into the breast milk of lactating rats and can form DNA adducts in the liver of the newborn with possible carcinogenic consequences (Ghoshal and Snyderwine, 1993). This suggests that such a route of exposure may also exist in other mammals, including humans.

There is little doubt that thermal processing is one of the most important techniques prolonging shelf life and improving quality and microbiological safety of foods. However, the technologies employed in the food industry should be designed in such a way that the possibility of mutagen formation is minimized.

MUTAGENS IN TEA, COFFEE, AND ALCOHOLIC BEVERAGES

Coffee brewed from roasted beans and that prepared from instant powder, including the caffeine-free type, all display mutagenic activity. Apart from natural mutagens such as caffeic acid and its precursors chlorogenic and neochlorogenic acids (Ames et al., 1990), these drinks contain mutagenic products of pyrolysis: methylglyoxal (Table 12.1) and less active glyoxal and diacetyl.

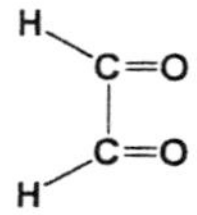

glyoxal 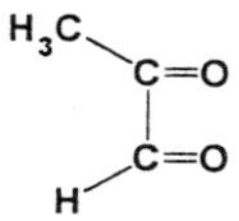   methylglyoxal 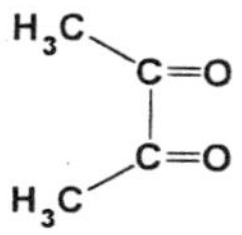   diacetyl

These pyrolysis products were also found in roasted tea, brandy-type alcoholic beverages, soy sauce, and soybean paste (Sugimura and Sato, 1983). In addition, as a result of ethanol metabolism, mutagenic acetaldehyde is formed, while in coffee and tea, caffeine is present, which is an inhibitor of DNA repair synthesis and may also contribute to cancer risk.

Although it is obvious to specialists that the beverages discussed contain many mutagens, they are accepted by the public for two reasons: (1) the benefits associated with their consumption are widely appreciated, and (2) their toxic constituents (ethanol, caffeine) are naturally occurring compounds. The same way of reasoning also concerns other food products that are considered safe only because they are of natural origin. It is not trivial to combat such common beliefs since hard evidence demonstrating unequivocally the involvement of certain food components in etiology of human cancer is very difficult to obtain because too many variables would have to be taken into account. Coffee is a very good example. It is estimated that a typical cup of coffee contains at least as much as 10 mg (40 ppm) of rodent carcinogens and 100 mg of caffeine. However, the epidemiological evidence to date is insufficient to show that coffee is a risk factor for cancer in humans (Ames and Gold, 1990).

FAT AND HIGH-CALORIE OR PROTEIN INTAKE

A number of epidemiological studies indicated that a high intake of fat contributes to the development of breast and large intestine cancers in humans (Ames, 1986). The effect of fat is exerted primarily after initiation of tumorigenesis.

Some constituents of fats, such as cholesterol and unsaturated fatty acids, are easily oxidized during thermal processing, giving rise to reactive molecules, which in turn, may trigger a chain reaction of lipid peroxidation leading to the formation of mutagens, promoters, and carcinogens. These include radicals; fatty acid epoxides and peroxides; aldehydes, e.g., malondialdehyde, which binds covalently with DNA; and others (Ames, 1986).

The high calorie and fat intake appears to be implicated in induction of endogenous oxidative damage of macromolecules, including the formation of so-called oxygen DNA adducts (e.g., 8-hydroxy-2′-deoxyguanosine) and protein carbonyl derivatives (Youngman et al., 1992). This type of lesions, caused by oxidants generated in large amounts during normal metabolism, are believed to play a significant role in the process of aging and the variety of degenerative age-related disorders, including cancer. The animal studies have shown that calorie and protein restriction markedly inhibits both carcinogenesis and accumulation of endogenous oxidative damage. Demonstrating the converse correlation is rather difficult since animals tend not to ingest excess calories (Youngman et al., 1992; Rogers et al., 1993). Some support, however, has been gained from the classic epidemiological study showing that breast cancer is the major neoplastic disease in the Western world of people on a high-fat diet, while with the traditional low-fat Japanese diet, postmenopausal breast cancer is relatively infrequent (Wynder et al., 1991).

ANTICARCINOGENIC FOOD COMPONENTS

A number of natural and synthetic compounds are able to prevent cancer induction and/or development when administered to animals before or concomitantly with carcinogens. These substances include vitamins, microelements, compounds of plant origin, medicines, and others (Table 12.3). Although the modes of action of anticarcinogens have not been fully understood yet, it is quite clear that any factor capable of counteracting the production of carcinogenic metabolites and inhibiting the initiation or promotion of tumorigenesis, as well as being able to prevent the formation of metastasis by malignant cells, may be considered an anticarcinogen.

Anticarcinogens are divided into three groups, depending on the stage of carcinogenesis at which they act (A Scientific Status Summary, 1993; Ames, 1986; Caragay, 1992). The first of these groups includes factors that prevent the formation of procarcinogens from precursors. For instance, vitamin C inhibits, via an unknown mechanism, the formation of carcino-

TABLE 12.3. Selected Categories of Chemical
Substances Displaying Anticarcinogenic Properties.

The Stage of Carcinogenesis Blocked	Category of Substances
Initiation	Organosulfides Terpenes Aromatic isothiocyanates Indoles Coumarins Phenols Dithiocarbamates Barbiturates
Promotion and Progression	Retinoids Protease inhibitors Terpenes Isothiocyanates Selenium salts

Source: Adapted from A Scientific Status Summary, 1993.

genic nitrosamines from amines and nitrites present in food (Caragay, 1992). To the second group belong the agents that protect cells against DNA damage. These mechanisms are best recognized, and they involve the reduction of synthesis or inhibition of enzymes responsible for metabolic activation of carcinogens and induction of enzymes taking part in detoxification of harmful substances. Also, agents capable of trapping DNA-damaging species belong to this group. The ability to modulate the activity of cytochrome P450 isoenzymes, often implicated in carcinogen activation, is displayed by numerous compounds, e.g., phenols, found in edible plants. The removal of toxic metabolites is accomplished, usually by nucleophilic substances, first of all glutathione, which can bind electrophilic DNA reactive intermediates. Vitamin E trapping oxygen radicals in lipid membranes, as well as β-carotene and other polypropenes, present in all chlorophyll-containing food products, are particularly effective in neutralization of singlet oxygen and protect DNA against oxidative damage. A similar role is played by selenium-containing compounds. Selenium is an essential component of the active site of glutathione peroxidase, the enzyme responsible for destroying hydrogen peroxide and other peroxides generated during lipid peroxidation. The factors belonging to the third group of anticarcinogens inhibit the neoplastic transformation of initiated (procancerous) cells. These mechanisms are least known. Such properties are displayed by retinoids, which inhibit both cancer promotion and formation of metastasis and help to restore cell–cell communication (Ames, 1986).

Vegetables and fruits are major sources of dietary anticarcinogens (Figure 12.1). Apart from the anticarcinogen types mentioned previously, edible plants and whole grain cereals contain fiber whose anticarcinogenic potential, especially in the case of colon and breast cancer, has been demonstrated in a number of epidemiological studies. Dietary fiber is thought to modify the biological actions of hormones with the effect of reducing the risk of hormone-related breast cancer. In addition, it increases fecal bulk, which results in dilution of bile acids exhibiting tumor-promoting activity, decrease of fecal mutagen concentration, and shortening of the transit time of fecal material, thus reducing the exposition of colonic mucosa on carcinogens (Greenwald and Clifford, 1994). It has also been found that lactic acid bacteria from both fermented dairy (Gilliland, 1990) and nondairy (Thyagaraja and Hosono, 1993) foods display antimutagenic activity and the ability to bind mutagens. In the binding, peptidoglycan present in the bacterial cell wall is involved, and this property is not abolished after sterilization.

The information presented here indicates that a diet rich in fruits and vegetables is the most recommended. Edible plants not only contain numerous anticarcinogenic substances, but also provide food of low calorie and protein content. All these factors reduce cancer risk in humans. Suggestions have been put forward that food products should be enriched with purified anticarcinogenic substances (Caragay, 1992). However, until the mechanisms of action of these substances are understood, such propositions cannot be put into practice.

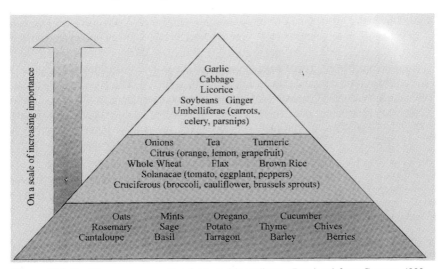

Figure 12.1 Possible cancer-preventive foods and ingredients. (Reprinted from Caragay, 1992, with permission.)

SUMMARY

Food contains potential carcinogens, tumor promoters, and anticarcinogens. These can either be natural constituents or may be formed from precursors found in a given food product. Among the techniques employed in the food industry, thermal processing is the major cause of formation of mutagenic substances. However, the risk associated with the generation of mutagens must be weighed against important benefits resulting from thermal processing, such as prolonged shelf life and microbiological safety of food. In addition, the amounts of mutagens produced during heating are very small compared to those inducing tumors in animals. For example, it is estimated that a person weighing 70 kg, eating about 200 g of fried beef and smoking twenty cigarettes, ingests roughly 3.5 μg of heterocyclic aromatic amines, 90% of which are derived from cigarette smoke. Induction of tumorigenesis in a rat weighing 0.1–0.5 kg requires that the animal is administered several milligrams of these compounds (A Scientific Status Summary, 1993).

It could then be concluded that the amounts of carcinogenic substances in food are so small that they should be readily detoxified within the organism. Food, however, is not the only source of mutagens and carcinogens in human surroundings. In the polluted environment, there is a plethora of factors that may increase cancer risk. Carcinogenic food components thus represent an additional burden, the one perhaps most effectively delivered into the human body. Therefore, the levels of these substances in food should be as small as possible. This refers to both the food industry, as well as every household. Modern technologies used in future food processing should ensure that, while food products retain the desirable properties, the formation of potential carcinogens is minimal. Equally important, from a reducing cancer risk point of view, are changes in cooking and dietary habits displayed by people. The latter are summarized by National Cancer Institute guidelines, which advocate reducing total fat, increasing consumption of dietary fiber and vegetables, avoiding obesity, and using alcohol only in moderation.

REFERENCES

Ames, B. N. 1986. "Food constituents as a source of mutagens, carcinogens, and anticarcinogens," in *Genetic Toxicology of the Diet*. I. Knudsen, ed. New York: Alan R. Liss, Inc., pp. 3–32.

Ames, B. N., Gold, L. S. 1990. "Chemical carcinogenesis: Too many rodent carcinogens," *Proc. Natl. Acad. Sci. U.S.A.*, 87:7772–7776.

Ames, B. N., McCann, J., Yamasaki, E. 1975. "Methods for detecting carcinogens and mutagens and the *Salmonella*/mammalian microsome mutagenicity test," *Mutat. Res.*, 31:347–364.

Ames, B. N., Profet, M., Gold, L. S. 1990. "Dietary pesticides (99.99% all natural)," *Proc. Natl. Acad. Sci. U.S.A.*, 87:7777–7781.

Caragay, A. B. 1992. "Cancer-preventive foods and ingredients," *Food Technol.*, 46(4):65–68.

Cassens, R. G. 1995. "Use of sodium nitrite in cured meats today," *Food Technol.*, 49(7):72–80, 115.

Dipple, A., Bigger, C. A. H. 1990. "Mechanism of action of food-associated polycyclic aromatic hydrocarbon carcinogenesis," *Mutat. Res.*, 259:263–276.

Fine, D. H., Challis, B. C., Hartman, P., Van Ryzin, J. 1982. "Endogenous syntheses of volatile nitrosamines: model calculations and risk assessment," in *N-Nitroso Compounds: Occurrence and Biological Effects*. H. Bartsch, I. K. O'Neill, M. Castegnaro, M. Okada, W. Davis, eds., Lyon, France: IARC, publ. no. 41.

Ghoshal, A., Snyderwine, E. 1993. "Excretion of food-derived heterocyclic amine carcinogens into breast milk of lactating rats and formation of DNA adducts in the newborn," *Carcinogenesis*, 14(11):2199–2203.

Gilliland, S. E. 1990. "Health and nutritional benefits from lactic acid bacteria," *FEMS Microbiol. Rev.*, 87:175–188.

Greenwald, P., Clifford, C. 1994. "Fiber and cancer: prevention research," in *Dietary Fiber in Health and Disease*. D. Kritschevsky, C. Bonfield, eds., St. Paul, Minnesota: Eagan Press, pp. 159–173.

Gross, G. A., Turesky, R. J., Fay, L. B., Stillwell, W. G., Skipper, P. L., Tannenbaum, S. R. 1993. "Heterocyclic aromatic amine formation in grilled bacon, bean and fish and in grill scrapings," *Carcinogenesis*, 14(11):2313–2318.

Havery, D. C., Fazio, T. 1985. "Human exposure to nitrosamines from foods," *Food Technol.*, 39(1):80–83.

Hecht, S. S. (1993). "Understanding carcinogens and anticarcinogens in food," *Food Technol.*, 47(1):15–16.

Hibbs Jr., J. B., Taintor, R. R., Vavrin, Z. 1987. "Macrophage cytotoxicity role for L-arginine deiminase and iminonitrogen oxidation to nitrite," *Science*, 235:473–481.

International Agency for Research on Cancer. 1987. "Overall evaluations of carcinogenicity: An updating of IARC Monographs, vol. 1 to 42," in *IARC Monographs on the Evaluation of Carcinogenic Risk to Humans*. Lyon, France: IARC, suppl. 7, pp. 83–87.

Krone, A. C., Yeh, S. M. J., Iwaoka, W. T. 1986. "Mutagen formation during commercial processing of foods," *Environ. Hlth. Perspect.*, 67:75–88.

Miller, A. J. 1985. "Processing-induced mutagens in muscle foods," *Food Technol.*, 39(2):75–79 & 109–113.

Pariza, M. W. 1982. "Mutagens in heated foods," *Food Technol.*, 36(3):53–56.

Preston-Martin, S., Pike, M. C., Ross, R. K., Jones, P. A., Henderson, B. E. 1990. "Increased cell division as a cause of human cancer," *Cancer Res.*, 50:7415–7421.

Rogers, A. E., Zeisel, S. H., Groopman, J. 1993. "Diet and carcinogenesis," *Carcinogenesis*, 14(11):2205–2217.

A Scientific Status Summary by the Institute of Food Technologists' Expert Panel on Food Safety & Nutrition. 1988. "The risk/benefit concept as applied to food," *Food Technol.*, 42(3):119–126.

A Scientific Status Summary by the Institute of Food Technologists' Expert Panel on Food Safety & Nutrition. 1993. "Potential mechanisms for food-related carcinogens and anticarcinogens," *Food Technol.*, 47(2):105–118.

Sikorski, Z. E. 1988. "Smoking of fish and carcinogens," in *Fish Smoking and Drying. The Effect of Smoking and Drying on the Nutritional Properties of Fish.* J. R. Burt, ed., Barking, England: Elsevier Applied Science, pp. 78–83.

Sugimura, T., Sato, S. 1983. "Mutagens-carcinogens in foods," *Cancer Res.* (suppl.), 43:2415s–2421s.

Swenberg, J. A., Richardson, F. C., Boucheron, J. A., Dyroff, M. C. 1985. "Relationship between DNA adduct formation and carcinogenesis," *Environ. Hlth. Perspect.*, 62:177–183.

Taylor, S. L. 1982. "Mutagenesis vs carcinogenesis," *Food Technol.*, 36(3):65–68 & 98–103.

Thyagaraja, N., Hosono, A. 1993. "Antimutagenicity of lactic acid bacteria from 'Idly' against food related mutagens," *J. Food Protect.*, 56:1061–1066.

Wakabayashi, K., Sugimura, T., Nagao, M. 1991. "Mutagens in foods," in *Genetic Toxicology,* A. P. Li, R. H. Heflich, eds., Caldwell: The Telford Press, pp. 303–338.

Wynder, El L., Fujita, W., Harris, R. E., Hirayama, T., Hiyama, T. 1991. "Comparative epidemiology of cancer between the United States and Japan: A second look," *Cancer,* 67:746–763.

Youngman, L. D., Park, J-Y. K., Ames, B. N. 1992. "Protein oxidation associated with aging is reduced by dietary restriction of protein or calories," *Proc. Natl. Acad. Sci. USA,* 89:9112–9116.

Index

285